NINA BURTON

NOTES FROM A SUMMER COTTAGE

The Intimate Life of the Outside World

Translation by
Rachel Willson-Broyles

MUDLARK

Mudlark
HarperCollins*Publishers*
1 London Bridge Street
London SE1 9GF

www.harpercollins.co.uk

HarperCollins*Publishers*
1st Floor, Watermarque Building, Ringsend Road
Dublin 4, Ireland

First published by Mudlark 2021
This edition published 2022

1 3 5 7 9 10 8 6 4 2

A catalogue record of this book is
available from the British Library

ISBN 978-0-00-846706-7

Printed and bound in the UK using 100% renewable
electricity at CPI Group (UK) Ltd

MIX
Paper from
responsible sources
FSC™ C007454
www.fsc.org

This book is produced from independently certified FSC™ paper
to ensure responsible forest management.

For more information visit: www.harpercollins.co.uk/green

Contents

Contents

Introduction

Into Nature

INVISIBLE AND FLOURISHING, fighting and loving – the many lives of Earth seethed around me. As a child I had registered my presence in the world by writing down my name, address and the specification 'Earth' in order to expand the walls around the centre of it all: me. I had questions when it turned out that other people, too, saw themselves as the focal point of the world. And, as if that wasn't enough, humans were not even the only protagonists – nature was full of them.

And what was nature? People said it was the environment and the outdoors, or the traits you were born with, but at the same time it seemed to have to do with endless birth, for 'nature' is akin to 'nativity'. In short, it was boundless life with billions of centres, all sparking with significance, each one moving to its own rhythms and perspectives until it was impossible to take it all in at once.

I studied the humanities course at upper-secondary school but took biology as an elective, and it was then, thanks to Linnaeus and Darwin sorting us among the animals, that I realised we humans belonged to nature. Later, I studied literature and philosophy at university, certain that this combination would lead me to answers about life. But literature mostly focused on individuals, and by that point philosophy was about abstractions. I longed to go back to the ancient Greek philosophers who posed questions about nature. Democritus wrote on atoms and stars; Thales knew all there was to know about water; Anaximander guessed, based on fossils, that we were distantly related to fish; and Heraclitus saw that everything shared the changeable nature of the rivers.

After them came Aristotle with his enthusiasm for every facet of life, from physics and meteorology to language and poetry. His interests were united in two Greek words: *bio* for life, and *logos* for word or reason. Each could be linked with other words, such as when they were combined to form 'biology'. Since he didn't want to live solely among theories, he retreated for a year to the island of Lesbos to study nature in a more concrete way. While his student Theophrastus sorted out the relation of plants to their environment, Aristotle devoted himself to the animals, and he charted their anatomy and development so meticulously that not only did he found the discipline of zoology – in many cases, his conclusions would hold water into our era.

While he started with 'the animal we know best', that is, humans, he later moved on to other species, for our greatness needed not diminish other creatures. He studied songbirds and doves, crows and woodpeckers, ants and bees, cephalopods and whales, foxes and other four-legged animals. He described the life cycle of the cicada and saw that snakes mated by winding around one another; dissected fertilised eggs and found that embryos already had eyes, veins and beating hearts. He wondered about heredity and suspected that it depended upon something he called *eidos*, imagining the concept as analogous to the order of letters in a word, and in doing so he came close to an explanation of hereditary DNA.

What was the driving force behind all of this life? Aristotle believed that each creature, while it lived, had a sort of soul that animated its matter and guided nutrients through its body. To him, it seemed nature had a unique ability to create ever more complex organisms, and since all of them must adapt to their environments, it was the environment that was the greatest determinant. It was like a household – arguments might arise, but there was still cooperation. Just like the sun and moon and stars, each part of the house had its role to play. All in all, this provided a context and a balanced framework for life, almost like the walls of a house, and, in fact, the Greek word for house, *oikos*, later gave us the word 'ecology'.

I was not a stranger to nature, even though I was a child of the city. We never had our own summer house, but during summer holidays Mum would rent us a country-side cottage somewhere or another, and a similar tradition continued once my sister got married abroad. She remedied her homesickness by renting Swedish summer homes, which I shared with her and the children before her husband's holiday time began.

For my own part, I spent about thirty years with successive live-apart partners who had their homes in the countryside. My interests had, in turn, been divided between them, for one, as an author, knew how words can broaden the world; the other, as a biologist, was familiar with the connections of nature. Like Doctor Dolittle, he gained the trust of animals and was even able to pet a capercaillie cock that had taken a liking to his porch. I myself mostly encountered the wild animals in the biologist's extensive library.

In other words, I had often been a guest in nature. I only became more than that after Mum's death, when we exchanged her apartment for a summer cottage. As in life itself, it was an inheritance that brought something new, and it manifested in different ways. For my sister, the cottage meant holidays with children and grandchildren, and for me it was a place to retreat with my manuscripts. After all, I wanted to write about nature and life. Could the cottage be the perfect place to do just that?

The cottage was situated on a large plot with a lively atmosphere. A small, mossy hill climbed from the south side, among pines and oaks, and to the west were hints of secret paths through blueberry thickets. To the north, the plot steeply abutted some public land, with the sparkling waters of a sound as a backdrop. The boundary lines were unmarked, so everything was at once private and open.

The property itself may have felt large, but the cottage was small in comparison. It was a single room that had been constructed on a whim in the traditional summer-cottage way and built out accordingly; the glass of the veranda replaced with walls to fit two bunk beds, and a kitchen and bathroom were added later in a small extension. After that, the topography prevented any further building.

Instead, the property had small outbuildings in each corner. In one was the former privy, transformed into a tool shed; in another was a carpentry shed next to an open storage shed. In the third corner, a small hut had served as a playhouse, and in the fourth corner was a small bunkhouse that I quietly designated a writing nook.

It was only natural for there to be certain shortcomings, especially given that the property had come with an exclusion clause. The carpenter who was brought in even muttered that it would be preferable to build new. How upsetting. Couldn't he see the idyll? What *did* he see?

Apparently, in any case, a number of repairs would be necessary. I was truly happy to deal with the craftsmen, for my books, too, feel like constructing buildings. Since the blueprints are always new, I must feel my way forward, and it's tricky to find the right proportions among different materials. So, I spend time at my desk working on problems of craftsmanship, on a daily basis.

I had a couple of projects that should be put to bed before I could devote myself to life and nature. One of them was about the way rivers run through both nature and culture, while the other was about how humanism in the Renaissance period united the humanities and the natural sciences. Erasmus of Rotterdam, who breathed fresh life into the essay genre, had become my hero, but I was also fascinated with the great encyclopaedist Conrad Gessner. Like Aristotle, Gessner was a student of half a dozen subjects, from zoology to philology. He wrote about thousands of plants and thousands of authors, and the relationships between animal species inspired him to examine the relationships between over a hundred languages.

I've always sympathised with the concept of encyclopaedias. It gives equal weight to both big and small, for there are no main characters – rather, it allows for a presentation of the world from various angles. To me, Gessner's perspective was echoed in the scope of life. I could devote only a few chapters to him in my Renaissance book, but I liked his methods of uniting creatures and languages, plants and literature.

The breadth of his seventy volumes would, of course, never fit in the tiny writing nook on the property, and the surroundings probably didn't accommodate many species either. And would I even comprehend their communication? What I knew about life on Earth had been transmitted to me by a human alphabet. The creatures that flew and crept, climbed and swam around me must have their own language, one that was suited for nature. They could literally be down-to-earth or light and winged, if they didn't venture forth tentatively like roots. So how would I discover animals with languages older than the alphabet? Differences generally erect walls between separate worlds.

But, as so often in life, this problem would provide its own solution.

Chapter One

The Blue Roof

YOU COULD SAY I GOT TO KNOW the cottage from above. The roof was the first thing the tradesmen looked at because the roofing felt needed to be replaced and the insulation supplemented. When an infrared camera was aimed up from inside the cottage the image turned the lavender-blue of a February evening. It indicated a massive amount of cold coming in. There was also the occasional small yellow blob among the blue, and since yellow meant heat there must have been remnants of insulation up there. The images got me to thinking. Here and there around the house lay tufts of insulation, like little fallen clouds. How had they got there? Surely they hadn't been blown out somehow?

The tradesmen would return in late March, and in order to meet them I slept over in the cottage. This was the first time I had done so, and there was still a winter chill, so as the radiators warmed up I went for a walk in the surrounding countryside. The light threw even the tiniest gravel pebble into shadows and relief on the bare ground where everything was ready to be adorned with life. A great tit warbled above a patch of coltsfoot, and lots of other things were surely taking form inside buds and cones bursting with seed. It felt as if a thousand discoveries awaited.

Back in the cottage, I made it just a little warmer by turning on the stove. As the water boiled for spaghetti, I dug through a few moving boxes from Mum's apartment. There was still a great deal to organise, but I planned to spend the evening relaxing and reading. The silence felt restful and was a good match for the book I'd brought along; it was about outer space.

After all, it was far out in space that the building blocks of life had once been born of a cosmos no greater in size than a fist. For one extraordinary second it was clasped hard around the coming galaxies and a limitless future. And then an eternal crescendo broke out. From a tangible beginning came a sky so brimful with stars, stars that spent some billions of years producing carbon and oxygen, silver and gold, and all the other ingredients necessary for life. Even the protons and electrons in my own body were once matter of radiation in space. All in all, then, I could consider myself a byproduct of dead stars, or perhaps a collection of their raw materials. There's plenty of it, for millions of tonnes of cosmic materials are still arriving on Earth.

I closed my eyes to think. From the perspective of the book I was reading, the Earth was part of an immense cycle of elementary particles that could be combined into rock, water, plants or animals. And as our fickle forms flickered by, our solar system made yet another circuit around the centre of the Milky Way. That circuit took between 225 and 250 million years and is called a galactic year.

Out in space, stars and planets moved like parts of a massive clockwork. Like all timepieces, it gets out of alignment from time to time, which explains why the moon is slowly distancing itself from us. For the present moment it wasn't changing much, as the distance only increases by four centimetres per year.

As the book painted me a picture of these proportions, space expanded the walls of the cottage. Even the smallest object, according to the astronomer author, could contribute to the bigger picture. If, for example, you were to hold a coin about a metre in front of your eyes, a hundred thousand galaxies could fit behind it, and each galaxy, in turn, would consist of hundreds of billion of stars. In our own Milky Way, the stars are scattered across such a vast area that the light from some of them has been travelling our way for millions of years. In the meantime, the stars themselves have died, but their light lives on in more or less the same way that old records carry on the music of dead musicians.

Where is that light going? Space has no centre point; it appears to be the same in every direction. I gave a melancholy thought to the probe that had been launched with a picture of two people onboard. Wasn't it a bit presumptuous to think of this as the most important information about Earth? And if there even was language in space, surely it was of a different character to our own. It was a world we had approached with mathematics rather than with words.

Perhaps a better introduction would have been NASA's recording of the electromagnetic vibration of Earth. It has been transformed into sound and when I heard this murmuring harmony with no beginning or end, I was moved in a particular way. Was this how we had imagined the music of the spheres? Kepler had speculated that Saturn and Jupiter were basses, while Earth and Venus were altos, Mars was a tenor and Mercury had the descant. I didn't know how they sounded in reality, but NASA's version of the song of Earth gave me a sense of our planet's life forms as both lovely and fragile.

Could I see any stars out there? I put down my book and went outside to stand there for a while with my jacket over my shoulders. According to the book I was reading, 90 per cent of the population of western Europe could no longer see a true starry sky, for the sky has been blotted out by our artificial lights. Certainly, space is dominated by darkness, but if we are, in fact, made of the stuff of stars, it might be fun to have a look at them. Only the North Star was faintly visible, twinkling through the atmosphere.

However, I could see something closer out of the corner of my eye. Wasn't that a shadow fluttering by? Were there bats here? I had mixed feelings about bats. They're the only mammal that has successfully conquered the air, and they are masterful flyers. Unlike birds, they have no feathers – just wings of bare skin stretched taut between the thumb and four fingers of their hands. The skin then extends all the way to the bones of their feet to give them a broader wingspan. And that's not all. Their winged hands manoeuvre faster in the air than my fingers can move on a computer keyboard.

For their part, bats communicate with rapid-fire ultrasound that probes the darkness where moths are foraging. Their personal communications, though, are more physical in nature and involve audible chattering. For instance, a female bat has been observed giving hands-on birthing assistance to a relative by first demonstrating how the body should be positioned to help the young come out more easily, and then receiving the young herself. It was not unlike a human birth. So why do warm, hairy bats seem so strange to us? Is it because we associate them with nighttime, when we retire and our senses are asleep?

After a while, I went in and lay down in one of the bunk beds. Although it was cramped, it also felt snug, as though someone else were in the top bunk. Warm bodies are protection against the desolate enormity and silence of outer space.

But suddenly I heard a sound very close by. Was that someone moving around on the roof above? I hardly thought it was a bat, so what was it? Since it was too dark to see anything outside I tried to fall asleep, although I longed for the morning light.

And when dawn broke, I wasn't the only one to awaken. I could hear the noise on the roof again; it was like tiny steps. Could it be a bird? When I sneaked out to check, it turned out there was nothing there. I did make a discovery behind the house, though. In the screening that covered the gap between roof and wall was a large hole. It looked like an entrance.

Throughout the day, this entrance would occupy my imagination as I tried to organise the moving boxes in the kitchen. Around lunchtime I made a circuit of the house and finally laid eyes on the mysterious roof creature. It was stretched out on the screening against the wall, dozing and enjoying what appeared to be a siesta. Its teeth indicated that it was a rodent, and at first glance it could have been mistaken for a rat. But its fuzzy tail said otherwise.

All at once, the pieces of the puzzle fell into place. This was a squirrel that had excavated the roof insulation to gain more living space, and clearly its ploy had worked. Judging from the infrared camera pictures, there must be a very grand squirrel apartment up there.

My emotions were thrown for a loop. Here was an intruder whose behaviour in relation to the house was highly unauthorised. On the other hand, I've always liked squirrels and was fairly knowledgeable about them, and now I could see both the sensitive whiskers on its forelegs and the rudimentary thumbs that made its front paws so much like hands. I looked at the fuzzy tail, which acts as a rudder in a leap between trees and as a blanket at night. Its softness was touching even without touching it.

Judging by the genitalia under its tail it was a female, and asocial female squirrels can have tough lives. After the hunt for a mate in the spring, they drive the male out of their territory and must care for all the young on their own. Just how hectic this life can be was driven home for me once when my biologist friend found a baby squirrel that had fallen out of the nest. I quickly read up on what squirrel mothers must do, which turned out to be an awful lot. The babies must be fed every three hours, and after that their tiny bellies should be licked or massaged to encourage digestion. Then, each one in turn must be dangled out of the nest for a while to keep it from becoming a latrine. It sounded like a full-time job, so I was doubly relieved when the squirrel mother found her baby. Maybe it had fallen out of the nest while she was trying to find a bit of food for herself between shifts. Her task would hardly become easier once the young began to dash around and become easy targets for hawks and cats, but female squirrels are so responsible that they even take on orphaned babies, if they're relatives.

My tender-heartedness was starting to take over. Squirrels have been hunted throughout almost all of history. They were sacrificed at Germanic festivals in winter and spring, and their tiny bodies provided the poor with food, as well as income from selling their pelts. In the 16th century, thirty thousand squirrel pelts might be exported from Stockholm in a single year, and that was just one of many stockpiles in Sweden. In more recent years, our native European red squirrels have been in competition with their grey relatives, who were brought here from the United States in the 20th century. The grey squirrels carry a virus that only they are immune to, and they sometimes create very tough little gangs that at times bite dogs and children.

This little red friend on the screening deserved protection. I gingerly sneaked away, and sat down quietly to read once I was inside again.

I had a little trouble concentrating on my book, because my thoughts kept being drawn back to my neighbour in the roof. What was it like to live with squirrels? Well, people have done it before. Ladies in both Antiquity and the Renaissance kept them as decorative pets. While it seems unlikely that they participated much in aristocratic society, an 18th-century English gentleman boasted about the musical abilities of his tame squirrels. They didn't care about choral music, but they could energetically stomp in time to chamber music from their cages. One squirrel had kept up an allegro rhythm for ten minutes before pausing and turning to another rhythm. In general, they probably found life indoors much less stimulating given the hamster wheels that were put in their cages.

At last the day became night again. By that point I really couldn't avoid thinking of the squirrel, because she was constantly moving around in the space below the roof. At first, I was amazed that we were only separated by a few planks. Hearing her movements gave me a sense of closeness, and I understood how bats can experience things without seeing them.

But soon enough, the fact that I could hear her went hand-in-hand with my being annoyed. Just as I dozed off, she started up again. Clearly, she was having a hard time falling asleep, and now so was I. It was like having a fussy child in the room. Each movement was a whine: something was in the wrong place, or maybe it was too hot. 'Go to sleep!' I hissed as she rummaged around up there. Squirrels aren't known for deploying any sort of interior decorating skills in their nests, but perhaps she was arranging what little insulation was left. If she'd used it for a bed, it must be too warm. Squirrel nests are typically lined with grass and moss, so I supposed mineral fibre might irritate her airways. In fact, wouldn't it be downright hazardous to her health?

The squirrel scratched herself loudly. She probably also had fleas. There's usually an awful lot of vermin in squirrel nests. I had previously had bad experiences with that sort of thing – one time, bird lice had spread to the vent above my bed. They came from pigeons in the attic, and I expected something similar could happen with squirrel fleas.

Now she was up and about again. Squirrels mark their territory by tramping around in their own urine and using their wet paws to stamp the boundaries. Was that what was going on up there? And didn't it sound like she was gnawing on something? Like other rodents, squirrels need to wear down their continuously growing front teeth every day.

After a restless night's sleep, I heard rustling from the roof around seven o'clock. Aha, the squirrel was awake. When I went to the kitchen I saw her peering in the window, probably on her way to breakfast.

As I drank my coffee I unearthed a pair of binoculars from the moving boxes so that I could spend time with her from a distance. To do so at closer range was impossible, because now a circus act had begun. As her kangaroo-like legs put a spring in her jumps, she wove every direction together like a dancing sunspot, here and there and up and down. I followed her movements with a jolt of nausea. It's possible for squirrels to leap five metres, and it's also possible for them to fall. But there was neither fear nor boldness in her jumps. They were made with breezy effort in a single, fluid *now*.

At last, she stopped in a spruce tree and I could focus the binoculars on her. She had found a breakfast cone. As her paws turned it in a spiral, she peeled it so systematically that the shell of a seed fell to the ground every four seconds. The whole cone was finished in seven minutes.

Then she vanished for a while as I got dressed and straightened up. When our paths crossed again at the corner of the cottage she greeted me with an annoyed flick of her tail. I felt a little hurt, given how thoughtful I had been towards her, but I suppose she had grown used to living undisturbed. Her quiet life wouldn't last much longer, though. During the night, I had made up my mind to be a troublesome neighbour. She must, like all squirrels, have several nests, and now it was time for her to choose a different one. The next time I heard her above me in the cottage, I banged loudly on the ceiling. Everything went quiet up there, so presumably she had got the message.

After all, indoors was not where I wished to encounter nature. During a circuit of the property I had heard a woodpecker drumming, and that seemed promising. They're said to thrive in forests with lots of biodiversity.

Not that I was out to spot anything rare. There are remarkable traits to be found even in the warbling great tit. I could no longer view great tits as just cute little birdies, because ever since it was discovered that they use tools and make plans, their intelligence has been considered on a par with that of chimpanzees. They hold pine needles in their beaks to pry caterpillars from cracks in trees, and they carefully note where other birds are hiding food in order to steal it. They can also send up a false alarm, a warning of nearby birds of prey, to scare competitors from the birdfeeder, and if they're truly hungry they can kill other small birds or sleeping bats. Peaceful individuals exist too, of course, so it's not as if the great tit has become one of Sweden's most common birds solely thanks to its cunning nature.

Suddenly I heard a more unexpected sound. Could it be, way out here? Yes indeed, I had just heard the most common bird in the world, a bird that exists in three times the number of all the humans on Earth. I had just heard a crowing rooster, so someone in the neighbourhood must have free-range chickens. It almost felt like the cosy idyll of a children's book.

Most chickens these days, of course, live far from nature. Hatched in incubators, the factory-farm laying hens are kept separately, while those raised for meat crowd among fifty thousand other chickens in windowless sheds – pre-emptively also fed antibiotics as a precaution against the crowded unsanitary conditions. Deep in the jungles of South East Asia, their wild forefathers still steal about in shy little flocks, so sensitive that they may die of shock if captured, just as hundreds of thousands of industrial hens do on their way to slaughter.

Jungle fowl were domesticated long ago in India, and Alexander the Great brought some back from his military expeditions there. For him, a chicken was a practical field supply that not only gave eggs and meat but also multiplied. In Greece and Rome, however, chickens would mostly be used for prophecies, since their manner of eating and taking flight were said to be signs one could interpret. Roosters were for their part viewed in a completely different fashion. If two aggressive individuals were placed in a so-called cockpit, from which neither could retreat, they would fight to the death. This popular spectacle would continue long into the 19th century in England, and names of breeds still live on in boxing terms like 'bantamweight'.

Hens, too, could demand respect outside the factories. This had become evident to me during the summer I rented a writing retreat cabin next to a henhouse, for the inhabitants strutted about freely during the day. Their patches of manure were about the size of a Danish pastry, and as I tried to keep from stepping in them I began to comprehend their internal hierarchy, from the top bird down to the bottom of the pecking order. That pattern certainly seemed familiar. Later I found out that their clucking involves more than thirty different sounds, including separate warning calls for airborne threats and those on the ground.

These were enormous hens who even managed to survive a fox attack, although their rooster didn't make it. After that, they were provided with a young cock who started out seemingly terrified of the massive ladies in his harem. Even the owners' youngest son was afraid of the hens, for he had heard that birds are descended from dinosaurs. You could tell just by looking at these giants.

The first person to suspect it was the biologist Thomas Henry Huxley. He was working on a dinosaur skeleton in 1868 when he was offered a turkey leg for dinner one evening, and he was struck by the similarities between the thigh bone on his plate and the one in his laboratory. Later, genetic analyses have proved him right. Chickens and turkeys truly are the closest relatives of dinosaurs. Perhaps the transformation began with small dinosaurs making their way into trees to avoid larger predators.

After all, chickens do still prefer to roost on a perch at night.

Soon the rooster stopped crowing, and after that all I could hear was a dove down in the public lands and a crow at the top of a conifer. I have to admit, I've never thought highly of either of these birds. Doves have become a symbol of peace and love and the holy spirit, but in real life they give rather a different impression. They were the ones who had given me bird lice once upon a time. And how had they come to be associated with the holy spirit? They're said to be related to the extinct dodo, whose Portuguese name, *doudo*, means 'stupid', because a small head doesn't exactly give the impression of genius. Nor do doves, who hardly seem to be aware of the eggs in their slapdash nests. But my image of them has changed thanks to more recent observations. Writer Jennifer Ackerman, for one, has collected a hefty documentation of intelligence on the wing.

Like chickens, doves have lived close to people longer than other birds, and we humans are the reason for their spread. Rock doves, or pigeons, were domesticated ten thousand years ago, about the same time as jungle fowl, because their tender young are a delicacy. Since rapid multiplication was the goal, males who were always eager to mate and females who produced many young were encouraged to breed. They had no problem living in close proximity to humans, and cities made perfect homes for them since cornices and balconies were much like the rock ledges of their original habitats.

In the 16th century, the Mughal emperor Akbar the Great kept over twenty thousand doves and bred them to promote desirable traits. This method of breeding later came to be practised here and there in Europe, and it also inspired Darwin's theory of evolution. If genes were so mutable that humans could select for various features among doves, then surely nature could do the same thing on a broader scale.

What dove breeders of the 19th century valued most was no longer the birds' meat but their phenomenal ability to find their way. This had made them letter carriers even in ancient Egypt and Rome, and they continued this messaging career until the advent of the telegraph. Networks of dovecotes were found not only at giants like Reuters News Agency and Rothschild & Co. bank, but were also used to pass on news in smaller contexts – the results of 19th-century Swedish sailing regattas were reported by a carrier pigeon who flew them to the printing press of the newspaper *Stockholms Dagblad*, where they were posted in a news window.

More serious tasks, too, were entrusted to pigeons. Explorers, spies and the military gave them assignments that could have been adapted into thrilling novels with winged heroes. In 1850, for instance, a pigeon flew four thousand kilometres in order to deliver a message from a polar expedition, although unfortunately the message itself disappeared along the way. During both world wars, all fighting parties used carrier pigeons, and some of them were even awarded medals of courage after being drawn into battle. One British pigeon stoically persevered in its task even after having part of a wing shot off. German pigeons had it no easier, for they faced both rifles and peregrine falcons.

Not only are pigeons courageous – they are quick and observant too. They can find their way thousands of miles over unfamiliar territory at a speed of 80 kilometres per hour, and as skilled observers they are unmatched. When pigeons were shown a series of landscape photographs taken in close succession, they were able to pinpoint differences that had escaped humans. The United States Coast Guard, as a result, trained them to find dots the colour of a typical life jacket, and then brought them by helicopter to areas where boats had capsized. They were able to pick out humans even among large waves.

Their visual talents became just as apparent in experiments of a more artistic nature. After a bit of training they could tell a work by Picasso from one by Monet and were able to differentiate Cubists such as Braque from Impressionists such as Renoir. Using signals for colours, patterns and textures, researchers were even able to get them to judge paintings as beautiful or ugly.

The list of their abilities doesn't stop there. It turns out pigeons are good at numbers and can place nine images of objects in the proper order. Their memory is also so exceptional that they can memorise a thousand images in one year and recognise them later when presented in negative or upside down.

Faced with such a tableau of winged intelligence, I felt ashamed of looking down on pigeons. After all, it's our fault they increased in number so rapidly and so strongly prefer to live near us, for those were the traits that we selected in breeding them. And it's certainly evident that they have lived alongside us for a long time. Not only can pigeons recognise individuals in their own flocks, they can also tell humans apart and identify expressions of emotion such as anger and sadness in photographs of different people.

Pigeon empathy is probably not the sole reason behind this. The ability to read emotions seems to be of some survival value. Birds can use it to detect threatening attacks and to join forces with each other by way of hardly perceptible signals. A look, a certain posture, a way of fluffing up feathers is all it takes. Even we humans unconsciously read others; in fact, tones of voice and facial expressions can be more honest than words. And, incidentally, words are said to transmit only about 7 per cent of what we communicate. Could it be that the art of reading between the lines is the basis of all communication?

Of course, here too I spotted a problem. It's easy to project emotions onto other creatures or slot them into neat patterns. Doves, for example, have solely been allowed to symbolise gentleness, while sharp-sightedness has been reserved for hawks. The cawing crow, for its part, has come to be seen as the opposite of the cooing dove, and when Ted Hughes deployed the crow in a collection of poems it became an anti-hero. Where the swift soared through the scent of violets, the crow chowed down on a dropped ice cream among the rubbish on the beach.

For how could the bawling of a crow inspire poetry? To me, the fact that they belong to the same order as songbirds seemed as incomprehensible as the fact that such categorisation has to do with the shape of their feet. The crow's biological relationship to birds of paradise was no less mysterious. That black and grey jacket of the hooded crow would be fitting for an undertaker, and its croak isn't particularly uplifting either.

But as we know, appearances can be deceiving. The Romans must have heard faith in the crow's song, for they interpreted 'kra kra' as *cras*, which means 'tomorrow' in Latin. So, to their ears, a caw sounded like eternal hope. And even I knew that crows really aren't all that dismal.

I once had the chance to come along on a weekend sailing trip to the outer archipelago. I brought two of my nephews; the female sailor brought her pet crow. She had previously asked whether we were afraid of crows, so I suppose she was used to people being suspicious of them.

During our voyage the crow mostly stood on deck, legs planted wide, steady like a seaman. While the sailor handled the craft, the crow seemed to believe its assignment was to watch over the passengers, as if in the discreet manner of a spy. My nephews were smokers at the time, and since they were constantly fiddling with their cigarettes they attracted the crow's attention.

We were allowed to choose our own sleeping arrangements once we reached the island where we would spend the night. One option was to share a cottage with the sailor and the crow, the other was to be on our own in the berths on the boat. We elected to stay on the boat in order to enjoy the rocking of the waves, and also because we weren't really used to sleeping near a crow. Apparently, it liked to spend the night sitting in an open doorway to keep an eye on everything.

After a night free of avian surveillance, one of the boys went up on deck for his first cigarette of the morning. But he'd hardly taken it out before the crow came swooping in from the cottage like a bat out of hell. She landed on his shoulder with a thud and began to thoroughly study his smoking, which now had a more serious bent. His cigarettes were running out.

At lunchtime, the smokers had a brotherly drawing of straws to see who would get the last cigarette, which was lit with great reverence. Suddenly the crow showed up again out of nowhere. She was coming straight for us, and in an acrobatic manoeuvre managed to snap up the cigarette and fly with it to the roof of the cabin. There she sat, taunting us, desirable object in her beak. It all became clear. She was no bad omen. She was simply having a lark.

Since then I had seen plenty of reports about the whims of crows. They play hide-and-seek with each other and tag with dogs. They tease cats. They catch sticks in the air.

They ride jar-lid sleds down snowy roofs, and when they get to the bottom they take the lid in their beak to do it all over again.

Playfulness can be the sister of creativity, and crows do their best to prove it. In one of his fables, Aesop tells of a thirsty crow who manages to reach the water at the bottom of a pitcher. She simply drops pebbles into it to raise the level of the water. In real-life experiments, crows have demonstrated the same behaviour, as well as that they can solve a number of problems that demand the use of tools.

In fact, crows seem to possess many of the traits we link with intelligence. They clearly have a sense of humour; they can plan; they are curious and adaptable and, at the same time, individualistic. Even back in ancient times they were drawn to the many opportunities of a city, although they did not allow themselves to be domesticated. Intelligence is also thought to be promoted by a prolonged childhood with instructive parents and a social life. Crows tick these boxes as well. Aristotle noticed that they cared for their young longer than other birds, and then maintained contact with family members. Now we know that they communicate by way of a multitude of sounds that differentiate not only species but individuals. What's more, they all seem to have their own identification sounds that others in their group recognise. They can even understand human body language, so they look in the right direction when someone points. Chimpanzees can't do that.

Like magpies, crows often gather around dead relatives, although it's not clear whether this is to serve as
witnesses to the death or to show loyalty. In any case, they
do have a good memory. It's a piece of cake for them to
match two identical images in the game of Memory. Their
eye for human faces is so good that the United States
Army tried to involve them in the hunt for Osama bin
Laden. They're especially skilled at recognising people
who have mistreated them, and can even teach other
crows to make out such villains from a distance. In
general, they keep an eye on everything around them, so
they, if anyone, would know what was going on around
the property.

It seemed a little awkward that the sharp-sighted creatures of the trees could perceive so much more about me
than I could about them. But I suppose that was the point.
Up in the trees, they could blend in with nature.

The squirrel was an exception. She may have preferred
to be left alone in her chosen nest, but she was hardly shy.
The next time I saw her come scampering by, I happened
to be eating an apple and tossed a piece her way. Although
she was in her usual hurry, she stopped to take me in. Not
that she stooped to being fed, but she did go so far as to
stand on her hind legs to get a better look at me. Her eyes
were large, like a child's, and her white belly gleamed
unguardedly. This time there were no annoyed flicks of her
tail. There and then, I decided to put up a squirrel feeder
full of nuts as compensation for confiscating her attic flat.

Having non-human company can be remarkably restful. From a psychological standpoint, it's not hard to see why. Such a relationship can be free of concepts like 'why', 'guilt' and 'forgiveness'. I could have given her lengthy explanations about roofing felt and insulation, and how the roof was the most important side of a house. I also could have laid out what might have happened if she'd stayed there under the roof. But there wouldn't have been any point. The grammar of the squirrel world was simpler than in my own world, without such complicated factors as a conditional 'could have been'. She wasn't well versed in cause and effect. The past was a memory of tangible seeds, tied to real places, and she sometimes forgot even that. When it came to possessive pronouns, all she needed was 'mine'.

Once the squirrel had apparently accepted her eviction, it felt as if one of the roof issues had been dealt with. And that was nice, since the carpenter and his helpers showed up after lunch. They wanted to tear up part of the roofing felt right away, to see how it looked underneath. A ladder was fetched and steadied, and two of the men entered the roof.

That's when it became clear that the squirrel and I were not on the same page at all. Quite the opposite – she took this as a brazen attack on her territory, and breaching of territories can unleash strong emotions. She came bolting through the trees like Tarzan on the warpath and, after a few intermediate stops, leapt onto the roof. There she stood on her hind legs and, from her full height, unleashed a tirade of chattering curses. She emphasised the gravity of the situation with several stomps of her foot. The tradesmen looked with an uneasy fascination, ducked back down, and began to remove the felt.

I have to admit, I found her impressive. What integrity, what courage! Still, she had lost this battle for territory, for the demolition was inevitable. Her attic apartment would be cleared out, and the house would belong to the tradesmen for a while. Once I had packed up my belongings and made one last circuit, I handed them the key.

By the time the new roofing felt and insulation were in place, spring had advanced a bit more and the roof work had reached a new stage. Gutters and downspouts would be installed, and a metalworker wanted to know if they should be fitted to water butts.

Naturally the house should have them, for water butts have both a practical function and cosy connotations. Thanks to the internet I managed to find a pair of used metal butts in the proper shade of green, and a seller who could transport them to the cottage. I just needed to be there to receive them. By that point I hoped I would also meet the migratory birds who made their summer home on the property. As a token housewarming present I had already purchased a birdhouse, and at the same time I bought a rain gauge to measure small amounts of precipitation.

The tradesmen were on holiday during the week I visited, but I still heard an eager hammering coming from the property. The woodpecker must have been inspired by the carpenters and the spring sunshine. It seems as if woodpeckers want to make new nesting holes each spring, where they can take turns brooding and feeding their young. Although they divide parenting up equitably, they aren't very social animals; yet their wild drumming helps them stick together and can probably transmit important information. In experiments, woodpeckers have even learned to ask for different objects by varying the number of beak-strikes. They are magnificent percussionists and change their tone by drumming against an assortment of materials.

Besides being a drumstick, their beak serves as hammer, lever, chisel and insect-detector, all in the span of one minute. This woodpecker was happy to demonstrate. When its gentle, exploratory taps located the larvae in a tree, it lifted the bark to remove the morsels, and its beak could then serve as a drill for a new nesting hole. I wondered how many holes it had already made on the property. A nuthatch lived in one of them. She had carefully narrowed the entrance with mud so the woodpecker wouldn't collect rent by eating up her young.

Like other homes, birds' nests can tell us something about their inhabitants, so while the woodpecker decorated with sawdust the nuthatch used pieces of bark. A blackbird nest that had fallen down near the storage shed was a piece of art. On the outside, braided conifer twigs had been plugged with moss and strands of birch bark, and then the inside had been smoothed over with mud and lined with soft grass. And why shouldn't birds be able to appreciate beauty and a well-made piece of craftwork? Some exotic breeds constantly redecorate their nests with fresh flowers, and others gather objects of a certain colour for installations where they dance for their ladies.

I felt like I had taken part in the birds' nesting work thanks to the birdhouse I had brought for them. Since I don't like climbing, I had chosen a model that could be hung from a branch. It was a little red house that looked rather like the cottage, but it wasn't actually a very good nesting site – it would probably be rough sailing in there when the wind blew. Perhaps this swaying little home could at least make a good birdfeeder when winter came.

Once I'd hung up the birdhouse, I devoted some time to the interior of my own house. There wasn't really much space to fill, but a lamp and Mum's meadow-flower curtains went up. Then I went back outside, and to my surprise I saw a blue tit fly straight into the failure of a birdhouse, without missing its swaying entrance hole. I heard only a plop as the little bird's body zoomed through the opening.

I was aware that blue tits aren't very particular about their nests, for in Stockholm one of them lived in my kitchen vent. If I leaned out of the window I could some-times see a pair of dark eyes peering back through the metal grate. We were a bit curious about each other. The first time she came flying by and saw me in the kitchen window, she quickly turned back to the elm in the neigh-bouring yard, where she had been collecting caterpillars.

As soon as I retreated into the room, she returned with her see-sawing flight path, but if I approached the window again, the same procedure repeated. We went back and forth a few times like that. The blue tit flew in and recoiled from my silhouette in the window; I backed up and she returned. It was like the steps of a dance. And as it went on, she became bolder. From within the room I soon saw how she sat on the window ledge to look for me. I'm sure the way she, like all birds, experienced the world differently to me contributed to her curiosity. For her, shapes and distances could partially be determined by way of shadows, and her gaze augmented objects slightly, all the better to spot insects. I must have been difficult to assess.

Our encounter was quite possibly made both safer and stranger since it took place through a windowpane. Author Björn von Rosen once described a nuthatch moving from window to window to follow his movements inside the house. Their contact was initiated when he fed her from the window ledge, and in time she also approached him outdoors. I never had such a relationship with the blue tit, but I had encountered a gaze that reflected a shy and yet curious individual.

She was one of the creatures who'd exchanged the shrinking deciduous forest for a city with very different demands. For birds, this didn't just mean the ability to make their nests in buildings rather than trees. Great tits must sing louder to be heard over the bustle of the city, and blackbirds' songs rise in tempo. In the loud urban environment, they wake up earlier and their biological clocks tick faster so they reach sexual maturity more quickly. It's almost as if they have been infected by the stress of the city. As urbanisation expanded, many animals were drawn into it, so from my balcony I could see a dozen species of birds. One day, too, feathers rained down over me on the pavement. A goshawk had found a pigeon nest, for one species had drawn the other along with it.

Still, it was not city walls I wanted to associate with birds, but the freedom of wings. On those occasions when I have felt a similar lightness at my desk, I have thought of all the feather quills that have been dipped in ink. For millennia they carried with them the ancient dream of moving freely, like Icarus and the angels, for even if a body was heavy, words and thoughts could take wing.

Leonardo da Vinci filled entire books with observations of birds in flight. He realised that air behaves more or less like water and noticed that it flowed both above and below a wing. Much later, with the help of his studies, the Wright Brothers were able to construct a flying machine. They understood, for instance, that a bird's tail feathers are a crucial factor in its ability to steer.

But no pilot has ever been able to measure up against birds. Everything about their flight astounded me. Some can stop abruptly and land on a swaying branch from a speed of 60 kilometres per hour, and some can sleep in the winds or mate in flight. Their feathers have almost become an extra sense – at their base, they transmit information about wind speed to the nerves of the skin. With the upward motion of the wings, the feathers spread – but as the wings push away the air, small barbs hold them together. No two feathers are identical, and none of them can manage without all the rest when it comes to flying.

It's not dissimilar to the way migratory birds stick together during their long journeys. They follow the same laws as so many things under the sun, for as the Earth moves around it, all life on the planet is affected. Happily, we don't notice that we're moving at a speed of 108,000 kilometres per hour, but the net of our meridians guides 50 billion migratory birds across the Earth. Some will fly many thousands of miles without a break in order to be close to the sun, and some even fly across the Himalayas. They may bring along seeds that have become stuck in their feathers and be accompanied by insects who are on a journey of their own. Their wings make the air vibrate as if in ecstasy. Up in the sky, millions of bird hearts are beating ten times faster than my own in order to pump vitality and warmth into bird bodies.

What drives all of these migrants? It's clear that they can feel differences in temperature, because millions of migratory birds have shortened their journeys as the climate has changed, and many of them have begun to stay put in the increasingly warm Nordic countries. Squirrels, too, have been able to adapt to changing temperatures. In Finland, it's been said since the Middle Ages that very hard winters drive them eastward. When this happens, they move as a group, in a front that is miles wide, although they mark their independence by maintaining a certain distance from one another. In Sweden, a similar squirrel migration was observed during the cold, snowy winter of 1955. Still, the most dramatic stories are those that came from Siberia, where throngs of squirrels set out on mass migrations, allowing neither mountains nor rivers to hinder them. Many were found exhausted, weeping sores on their paws, sometimes paralysed, but each one that could go on did. In the bitterly cold year of 1847, thousands of squirrels swam across the Yenisey River, only to crowd into the city of Krasnoyarsk, where they were killed en masse.

This phenomenon prompted me to spend a little more time wondering about squirrels. They're territorial loners, so what could force them so communally eastwards? Did they influence one another after all, or did they have some inner barometer that sensed such a strong temperature shift approaching?

For their part, migratory birds do appear to have internal barometers and light meters. As soon as the light begins to dim in an autumn locale, billions of birds suddenly head south like a giant flock of charter tourists. Like all airline passengers they must avoid overweight luggage, so they need to know, down to the gram, how much food they can store in their bodies. In many cases, the calorific content of a nut is enough for a journey to Africa. After all, other things with mass are needed in the body. The chest muscles that are required for wingbeats grow quickly, and so do brain cells that help them find their way. Bladders, however, are ballast that evolution has done away with, since any waste can be jettisoned during the trip.

The timetable is long since imprinted, and no one wants to be left behind. In 1933, a German bird station was taking care of a stork with a wounded wing, and he was struck by such a great urgency to migrate that he escaped. He couldn't fly, but he spent six autumn weeks walking 150 kilometres on foot in the same direction as his relatives had flown. This course had most likely been fixed inside him since his chickhood, because even first-time migrators must find their way. A similar anxiety beset some captive starlings. Facing south, they stubbornly flapped their wings against the bars of their cage all through migration season.

Migratory birds also have internal maps. These provide not only an image of the Earth but also the braille of the stars. Tern chicks stare at the sky while in the nest, and after just a few weeks they have memorised the position of the sun and various stars. The Pole Star is their north. Before leaving the nest, they take an extra lap around it to imprint the local geography. During their journey, it will expand infinitely.

It's this very thing, the journey of migrating birds, that brings a geography lesson to life in *The Wonderful Adventures of Nils*. This children's novel was meant to be a textbook for schools, and the publisher had helpfully supplied author Selma Lagerlöf with the dry bundles of facts used by headmasters. She read them all with fading enthusiasm. How could she make a textbook vivid? How could she breathe life into topographical fields of green-gold, and paragraphs about climate and flora? Her solution was to send animals into the landscape, and suddenly there was movement in the impenetrable thicket while songs were heard in the trees.

The idea that animals should drive the plot came from Rudyard Kipling's *The Jungle Book*, where a boy learns the languages and morals of creatures. Kipling's alpha wolf Akela is paralleled in Lagerlöf's head goose Akka, and Kipling's tiger Shere Khan was kin to her Smirre Fox. Since the flora and fauna of the jungle are unlike Sweden's, the landscape was reflected in moose, mallards, swans and eagles. They acted more or less as they do in nature, and although they were allowed to speak they were not humanised as in the old fables or in the Disney worlds that would come later. They merely demonstrated that humans were not the be all and end all of life on Earth.

It was unrealistic for Smirre Fox to chase a flock of geese all across Sweden, of course, but he became the common thread that gave purpose and excitement to the wild geese's fairly rambling journey. Lagerlöf was not unfamiliar with the natural sciences. She knew from her childhood home, for example, that a domesticated goose could run off with wild geese and return with its own young, and she did consult an expert in migratory birds on the behaviour of the wild geese. At the same time, it was through language that she brought nature to life, and author Michel Tournier would later place *The Wonderful*

Adventures of Nils among classics like the fables of Fontaine and Saint-Exupéry's *The Little Prince*.

For me, Nils Holgersson's winged friends outshone the most unlikely of inventions in the story. Birds were like a fairy tale in real life as well. Light as letters, and thanks to their incredible senses, they could make their way across stormy seas and vast continents to the exact spot they wanted to reach.

I once had the desire to observe their flight at close range, so one warm September evening I caught the last bus near Vemmenhög, where Nils Holgersson had begun his journey. I had brought along a small tent and a constellation umbrella so I could orienteer like the birds. My destination was the Falsterbo coast, where they would pass over. It was already dark when I arrived at the final stop, but guided by the lighthouse beacon I found a place where the grass was short and the horizon near. Once I was settled, I heard a gentle murmur above the tent. I'd seen pictures of how millions of migrating birds can cover the entire coastline on a radar screen, until it appeared to bloom and slip out into the sea. Now that flying coastline was above me.

I lay under that murmur, nestled in the down of my sleeping bag like the wing of a wild goose. I imagined lakes, rivers and mountains flowing into the flying birds' memories faster than into any schoolchild's mind. After all, birds must be able to match latitude to longitude no matter where they are. Their senses were with them, and their view was broad, thanks to the placement of their eyes on either side of their heads. Across the sea, too, came infrasound from the movement of the waves, but the most important part was their sense of the Earth. Deep inside our planet are rivers of glowing iron that create magnetic fields to which the birds orient themselves like iron filings. These gave their flight a clear direction, disturbed only by electromagnetism from the electrical apparatus of cities. They were living in much more intense contact with the Earth and the sun than we do.

When I peered out of the tent at dawn, I was still in the borderland between the escape of dreams and the facts of reality. A fairy tale egg gleamed in the grass. Upon closer inspection, this discovery only highlighted the extent to which the birds' ability to understand their surroundings exceeded my own, for the egg was a ball and my nocturnal territory a golf course. But once I packed up my tent, I took with me the thought that even an egg contains everything that might develop into a journey that spans the world.

An egg is the bird's world of origin, so if you want to make a connection with the tiny life contained within the shell, you must start there. Konrad Lorenz did just that with the greylag geese he studied. He had been fascinated with them since childhood, when he heard a flock of wild geese passing by the River Danube. Without knowing where they were going, he, in a burst of youthful longing, wanted to join them. He later tried to express his feelings in images, and they all depicted geese.

As a zoologist he would come to follow their lives in a different way. His house was already full of aquarium fish, dogs, primates, rodents, parrots and jackdaws, but he came to have a special relationship with the greylag geese he raised. To see how they emerged from their eggs, he let a domesticated goose brood a few of them until they were about to hatch. Then he moved one to an egg-hatching apparatus so he could attend the birth. When he put his ear to the egg, he heard cheeping, knocking and rustling inside. Then a hole appeared in the shell and a beak stuck out, and after a while an eye met his own. Then came the greylag gosling's contact call, a tiny whisper, and he imitated it in response. With this little greeting ceremony he had become a greylag parent, for now the chick had imprinted on him.

This wasn't something he meant to do, but forever after it became impossible for him to leave the chick for even a moment. Each time he tried to do so it let out a heartrending peep, so he had to carry it around in a basket during the day and bring it into his bed at night. At regular intervals came the contact call, an inquiring little *vivivivi*? In *The Wonderful Adventures of Nils*, Selma Lagerlöf interpreted this call to mean, 'Here I am, where are you?' and Lorenz felt this was accurate. He must continuously keep up the conversation throughout the chick's upbringing, since helpless baby birds need constant contact. When the gosling and her siblings grew up, he would take them on walks to meadows where they could eat fresh leaves, or to lakes where they could swim together, and when the birds flew he ran beneath them with his arms spread wide. To get them to land, all he had to do was fall into a crouch.

But geese must also communicate with one another, not least on long flights. They keep tabs on each other by way of a noisy conversation even as they stick close together like cyclists in a race. Like cranes, they fly in a V-formation so each bird's wing creates a spiral of air that helps lift the birds behind them.

Other birds form different sorts of groups when they fly. Further south I had seen tens of thousands of starlings moving in shapes that flowed into and out of each other like clouds, or like abstract figures in an ever-changing web. As though made of breath they would rise and sink from the horizon, the cloud contracting and expanding. One moment they seemed to be a fingerprint in the air; the next, a floating gas. These flocks are called murmurations. I liked that word. It's related to words like murmur, hum and buzz – when individual voices blend into a greater whole.

But how did such flocks form? People were beginning to understand how each individual in them keeps track of every other, at least. Since birds have a wider field of vision and faster reaction times than humans, each bird can keep an eye on seven others. Still, their lightning-quick coordination remained a mystery. Even when hundreds of thousands of them flew in tight formation, they never collided. They could maintain speed and change direction in seven-tenths of a second, and no typical form of communication is possible at such speed. What's more, each individual had a slower reaction time. Did they have some kind of invisible direct contact?

Yes, in fact they did. In the 1990s, it was discovered that there are special nerve cells in the brain that can transmit impulses for behaviours seen in others. These are called mirror neurons, and they're why laughter, gestures and yawns can be infectious. They can also transmit barely noticeable movements in flocks of birds, because in social groups it's important for each member to be in tune with others.

Did that mean each reaction perhaps even amplified the next? I thought of a phenomenon called the 'hundredth monkey effect', from a study of monkeys on an island in Japan. The researchers fed them sweet potatoes, and one day a young female had an idea. She began to wash the sweet potatoes in the sea to remove the dirt, and gradually others copied her. Then something remarkable happened. Let's say that it began when a hundred monkeys had adopted the behaviour. Suddenly, other monkeys on nearby islands also began to wash sweet potatoes.

Around the same time, something similar had been observed among birds. In the 1950s, milk bottles in England came with flimsy aluminium tops. The milk was delivered in front of each house every morning, and soon the blue tits of London learned that they could peck their way through the tops to reach the uppermost layer of cream inside. It didn't take long for all the blue tits of England to learn the trick.

It seemed as if, when an adopted behaviour reached a certain level of uptake, groups could both develop faster and change character, almost as if they had reached some critical mass. In his book of essays entitled *Crowds and Power*, Elias Canetti described how a flock of people could suddenly transform into a mob. Similarly, they could be swept up by ideas and cultural movements. I myself had seen how poets could change course as a group, like starlings, even though their poetry was about individuality. I had even written a book about it.

There was something about group psychology that I found both gripping and disturbing. It reminded me of two dreams I'd had as a child. In one, I was flying free with my arms stretched out and my legs tucked beneath me, an archetypical fantasy of flight. In the other dream, however, I was watching some sort of aliens inject people with a similarity serum. Everyone who received an injection wanted to convince me that the transformation felt good, but to me the encroaching uniformity was a nightmare. I didn't know if what scared me was the regimentation or the loss of control. I just knew that, like the squirrel, I wanted to preserve my integrity, and, like the birds, I wanted to fly free.

But the question was, of course, just how free they could be.

There is a dynamic at play between freedom and unity, solitude and togetherness, and birds provide great evidence of it. In the midst of their winged existence they are marked by the changing seasons, the stamp of their environment and the genes of past generations. They also look to each other for wayfinding and protection. On the cliffs at Stora Karlsö I had seen thousands of guillemots crowding close together, the better to avoid being taken by a bird of prey, and as their neighbours dived into the sea they also indicated where the fish were.

At the same time, they were all unique, and every bird could find its own egg among fourteen thousand others. Each newly hatched chick could also recognise its parents' calls amid the deafening racket. It had heard them through the shell before hatching and could pick them out among thousands of others.

Everywhere, the same phenomenon. Hardly noticeable nuances set each egg, each bird, each song apart, for life is borne upon a billion existences with fates beyond all classification. It wasn't visible in the migrants' autumn formations in the sky, but it certainly made up for it when they returned in the spring. At that point the group cohesion from the journey had vanished and their communication changed character. The same air that had carried flocks of birds was now full of territorial songs meant to keep males of the same species away.

For the song was not just about species. It combined an 'I' and a 'we' – a first and last name of sorts – and that 'I' now had to blossom with tiny nuances that would prompt a female to choose one singer over others. Then his own personal nature might eventually introduce more variations into the entire species – into the 'we'.

It touched my heart to hear this choir of birdsong, with each one blaring out his own fragile ego. Even the modest trill of the blue tit thought itself to be the centre of the world. Then again, why not? The eternal crescendo of the Big Bang, after all, began as a fist-sized centre of all possibility. So why couldn't a few simple tones also have meaning?

Does a song say more than 'Here I am'? I'd learned to identify birds' songs with the help of mnemonics. In these Swedish phrasebooks, the yellowhammer, for instance, spoke English: A piece of bread and no cheese. These little jingles provided a rhythm but hardly the meaning, nor was the sound really captured by the combinations of letters cited. The blackbird's song is definitely something other than true truly true tee tee.

Birds communicate with tones, of course, and we belong to a relatively tone-deaf species. We can't hear overtones or the 750 sounds a wren can utter in one minute. Even the chaffinch's song has subtleties we can only hear when a recording is played ten times slower than normal.

What's more, birds have a different sort of throat than we do; their syrinx can produce more than one tone at a time. In Greek mythology, Syrinx was a nymph who escaped the lustful Pan by transforming into some reeds. They sang when he huffed breath across them in frustration, and he cut them to make a set of pan pipes that could play several notes at once. A bird's syrinx can create rapid-fire, complex sounds in a single breath, with no audible break, thanks to the air sacs that take up a third of their body and their respiration rate of twenty breaths per second.

It seems that beauty is also important to birds. When they sing successfully, their bodies can even reap a chemical reward in the form of dopamine and oxytocin. This is especially true when they sing their autumn songs, which are not meant to mark territory or attract mates but only exist for their own sake.

And they're not the only ones to enjoy these sounds, of course. When my own forefathers, a hundred thousand years ago, developed throats that could shape a voice, it was birdsong they wanted to imitate. The oldest known instruments are flutes that were often made of bird bones. Later, when words arrived, only poetry could produce something related to the tone, rhythm and resonance of song, for poetry has roots in music. Greek poetry was originally set to music, so Aristotle found the rising tone of iambs fit for a dance.

So what did Aristotle think as he walked around Lesbos listening to birds? Did he think they captured something of what he had written about, such as poetry, the heavens, the soul and the brevity of life? Perhaps he compared birdsong to his own Poetics, as if in a grammar of musical notes?

I would have loved to discuss this with him. He would surely have been interested in what researchers have discovered about the effects of tone. For instance, the cries one might hear from dog owners and parents of small children alike can express different things in two notes. Abrupt tones can serve as a warning or admonishment when they fall (*Bad boy!*), but if they rise they become an order (*Come here!*). Longer, gentle tones urge calm when they fall (*There, there!*), but are encouraging when they rise (*Well done!*). Emotion is carried in the tone of voice, and it can even speak to those who do not understand the words. Perhaps the rhythm, too, sends a subconscious reminder of what the mother's heartbeat sounded like from inside the uterus. When she was calm, the heart beat slowly, and when she was upset or tense, it grew faster.

Aristotle truly wanted to understand the avian world. While his student Theophrastus immersed himself in lilies and marjoram, he pored over 140 species of birds, from the form and function of their beaks to the shades of colour in the egg and yolk. But, above all, he wanted to immerse himself in the life of a bird. He was the first who attempted to explain their annual migration, and he came to an astounding number of realisations about birdsong.

He noticed, for instance, that a bird didn't have its song straight from the egg, but had to learn it. Today, we know this is the case. When chicks in a nest hear their father singing, networks of nerve cells blossom in their brains. If they don't have a teacher, their song is unrecognisable. They must practise their melodies tens of thousands of times, always comparing with the memory of their father's song. Even so, the final product can sport small individual traits.

Even humans have taught chicks to sing. In the 19th century, German foresters stole young bullfinches from the nest and persistently whistled melodies as they fed the babies. And surprisingly enough, the young adopted those whistled tunes, even though they might be fragments of folk songs or a bit of some classical piece, and even though bullfinches aren't generally big singers. Starlings are even better at imitating melodies; in his time, Mozart had a tame starling that learned to whistle the theme from one of his piano sonatas.

The best imitators, though, are parrots. In the wild they're very social and communicative, and among humans they can copy not only melodies and instruments but also tone of voice and sentences. It seems Aristotle had some experience with them, in fact, because he pointed out that alcohol made them unusually cheeky. Had he perhaps shared a glass of retsina with a parrot on Lesbos? Even if it wasn't exactly an exchange of ideas on par with the Academy at Athens, he was sensitive to the fact that other species could have a language. In this, too, he was a pioneer.

The gift for language is most evident in the grey parrot – according to *The Guinness Book of World Records*, one of them mastered eight hundred words. Still, the most famous example is the grey parrot named Alex. Researcher Irene Pepperberg taught him a basic version of English for a very specific reason. She wanted to demonstrate that birds could understand abstract concepts and complex questions.

Since birds lack lips, Alex had a hard time pronouncing 'p', but soon enough he had mastered about a hundred words that could be used to test him. He had no trouble identifying fifty objects, seven colours, five shapes and various materials. He understood numbers up to six, as well as 'zero' or 'none'. He could differentiate between concepts like 'bigger' and 'smaller', 'the same' and 'different'. He could also express emotions, and when he didn't want something he gave a firm 'no'. When he did want something, he could use words creatively. He called an apple a 'banerry', because it tasted like a banana but looked like a cherry. To describe a cake, he invented the word 'yummy-bread'.

He had learned words by observing two research assistants that sat in front of him and asked each other for various objects. To help him better identify with them, they tried to sit in bird-like positions, but Alex even adopted the researchers' replies meant only for each other and used them in the proper way. Although human words are not birds' speciality, they did give some insight into Alex's sensitive brain. He successfully showed that birds can comprehend abstract concepts and complex questions.

Indeed, it was clear that humans had missed out on quite a bit in the mental world of birds. We had underestimated their intelligence as well as their communication. It wasn't just that many of the tones they produce are beyond our range of hearing. Even the internal order of the sounds seems to carry meaning. Black-capped chickadees can

combine six tones in various ways, more or less like sylla-
bles of different words.

Did that mean that birdsong can be compared to our
speech? This was a possibility both Aristotle and Darwin
were open to. And it does, in fact, turn out that there is a
connection, although it is hidden deep down inside birds'
brains. When researchers were obsessed with measuring
skulls, birds fell near the bottom of the scale of intelligence. When attention was directed inward, though, into
the nerve cells within the brain, those of humans and
birds turned out to be linked in a similar fashion, and it
was found that learning takes place in more or less the
same areas of the brain. The difference is that birds'
neurons are packed tightly, with quick connections,
because there is less space in their brains.

And that wasn't all. It was also discovered that there
are common genetics behind the similarities. A gene
discovered in 1998 was given the awkward name
Forkhead box protein P2, typically abbreviated as
FOXP2. It became popularly known as the language gene,
because mutations can cause language impairments and
possibly even autism. But we are not alone in possessing
this gene. It is also found in other animals, where mutations cause issues similar to ours – so, in birds, such mutations lead to stuttering or difficulty imitating sounds.

High above me, a blackbird was singing its way up a
pine tree, without even the tiniest stutter. Its neurons
communicated quick as lightning, and if song is a
language, the blackbird was a linguistic genius. In general,

they're probably no brighter than the monotonous dove or the inarticulate crow, for intelligence can also be silent, and there is endless variety in languages. But the blackbird's song may well be the loveliest of all. The fact that it arrives in the spring and then falls silent gives it a transitory intensiveness. Just as with our own songs, it probably has to do with good old love, and yet each bird adds just a bit of personal flair to its song, and the age-old themes of life and poetry continue.

As it turned out, there weren't many migratory birds for me to welcome, for most of the birds seemed to have overwintered on the property. At last, though, I did have one arrival – the water butts that would collect rainwater from the roof. After they were noisily rolled to the corners of the cottage, I invited the seller to have a cup of inaugural coffee beside them, and when I thanked him for the delivery he told me that the barrels were quite welltravelled. They had previously transported juice from some exotic country to the harbour in Rotterdam, and from there had made their way to a grocer in southern Sweden. In my mind, I associated their journey with the world of migrating birds. Where the birds had followed the shining sun, the barrels had mirrored their journey full of golden juice. As I heard a couple of gulls calling from the sound in the distance, I even felt a hint of the harbour atmosphere from the barrels' stopover.

It was in that frame of mind I wanted to eat dinner once I was alone again. As the fish au gratin I had brought was heating, I wiped down the patio table I had inherited. Some bird with no sense of table manners had just left a calling card on it, so I put down a tablecloth before I brought out the meal. Appetising steam rose from it, and all I needed to complete the tableau was a cold beer. It took only thirty seconds to fetch one, but that was just enough time for an attentive bird to make its move. When I came back outside, a common gull was standing in the middle of the mashed potatoes.

This sneak attack had come out of the blue. I hadn't seen any gulls around, but I suppose they must have noticed the smell of fish all the way out in the sound. The gull took off, contented, its feet covered in sauce. The fillets of fish had already been devoured.

As a teenager, I loved gulls for the way they seemed to float above the sea. I didn't know much about them back then and, apparently, I shared my uninformed love with many others – in the 1970s, the story of Jonathan Livingston Seagull sold a million copies and was even made into a film. It was about a philosophical aerial artist who soared in solitude above the materialistic squabbles of his flock. He didn't have much in common with actual gulls, who are extremely social. When zoologist Niko Tinbergen began to study them in the 1950s, he found an entire society opening before him. Each movement of the body and each sound of the voice relayed information about food and danger, anger and submission, cooperation and mate finding, chick rearing and suitable nesting sites.

Like so many birds, gulls have been drawn to developed areas where rubbish heaps and eating establishments always have food to offer. The roofs of buildings make safer homes than coastland, so from my apartment in Stockholm I had been able to follow the fates of a gull family on the next roof over. I watched the babies learn to fly, and when one chick fell down, the gull mother torpedoed herself at every approaching pedestrian on the pavement.

Gulls move as easily through varied environments as they do in the air. They can drink both saltwater and freshwater, and their menu runs the gamut from fish to small rodents and all the edible tidbits humans scatter around. They're inventive, too; they lure worms out of the ground by performing an energetic stomping that mimics the sound of rain. They've even been observed attracting goldfish in ponds by holding bits of bread in their beaks. For a clever gull, there are a thousand dinner options. And why should a gull see any difference between a fish in the water and a factory-made fish au gratin? At least I still had my beer and could make a sandwich to accompany it.

It was still light outside. Above me, the blackbird transformed air into an endless variety of songs, for the sky has no boundaries. High above the ground it is full of life, so as my dinner rested half-digested between soaring wings, I felt every breath of air I took was shared with thousands of others.

Yes, even with the squirrel, who was just scampering across the very roof where she had so recently made a scene. Now she seemed satisfied once more, and that made me happy too. Although it was odd how she so suddenly vanished by the roof. When I went over to the cottage, I had a sense of déjà vu. The roof was deserted, and the new screening the tradesmen had installed between roof and wall was a nice, fresh, springy green. But there was something familiar about it: a new hole had very recently been chewed in the same corner where the squirrel had made her old entrance.

Chapter Two

Wingbeats at the Door

THE BIRDS WEREN'T THE ONLY ONES having a hectic spring. The tradesmen and I had any number of tasks to finish up around the cottage before summer arrived, so my trips there became more frequent. The squirrel was so disturbed by my presence that she no longer felt comfortable in the house, and in fact this was part of my plan. But I also enjoyed visiting the property while the year was unfolding into life. The birds were singing, and as buds emerged the insects began to awaken. The light on tiny wings was not grand or flaring, but it was still intense.

As early as March, I found a sleepy fly bumbling around in a window. As I waved it outside, I thought of the massive number of insects it takes to feed a family of great tits. If that fly found a partner before it was eaten up, there could be as many as a hundred thousand new flies within the month – so, out on a mating adventure, fly!

A little later I had to help a newly awakened brimstone butterfly out of the rain gauge. It was a sunny yellow male who must have been eager to emerge from his hibernation spot so he would be ready when the females woke up. Springtime feelings were clearly not only for the birds. Butterflies, too, have hearts that beat faster at the scent of a potential partner, and it seemed to me that this yearning was especially fervent in brimstones. Their mating period can last for a week, as the male truly wants to give his female everything, including nutrients and hormones that increase her rate of egg-laying. Which meant that perhaps soon I could find her tiny eggs on select leaves.

Before I realised that thirsty insects need a dish of water, the narrow rain gauge inadvertently became a bug trap. The next one to take a dip in it was a big bumblebee. She was so tousled and tired by the time I fished her out that I went in to get a spoonful of sugar water. My rescue efforts were clearly appreciated. As she dipped her trunk-like proboscis into the spoon, I thought I could see her low spirits melt away. She fluffed up her fuzzy coat with some acrobatic assistance from her legs, and as its strands began to gleam in the sunlight I felt the urge to stroke them with my finger.

I knew how soft a bumblebee's fur could be, because I had once felt it against my skin. This was on a hot summer bus trip during which a bumblebee kept stubbornly circling me. Perhaps I was wearing some floral perfume, but her attention became so intimate that a man in the seat next to mine felt the chivalrous urge to shoo her away. Instead he managed to wave her into my décolletage. It would have been less than chivalrous to remove the bee from her new location, so there she stayed.

There was a gentle tickle as she moved around. She didn't sting me, because I had bent over to keep from squishing her, and anyway, 'she' might have been a 'he'. Since the stinger develops from an ovipositor, only females have one, and they don't like to use it unless they have to. Instead, they might start by lifting one leg in warning, or coughing up an unpleasant odour of butyric acid.

This bumblebee seemed to accept her new company, so I could only return the favour. I would have reacted differently had I been dealing with an earwig. It's not fair to the earwig, but insects have their skeletons outside their bodies, and bare skeletons arouse unpleasant associations. It's a different story if they have colourful forewings like ladybirds or hair like bumblebees. And bumblebees are incredibly hairy – American researchers claim to have counted up to three million tiny hairs, which is as many as a squirrel has. This struck me as an improbable amount, but pressed against my skin it did feel reliably soft. During the long journey, as the bumblebee rested against my skin, I made up my mind to learn all I could about my travel companion.

My amateur interest in biology had shifted as time went by. As a child I was fascinated by exotic mammals such as the shy okapi. It's a bizarre cross between giraffe, zebra and antelope, and appropriately enough, it even has similarities with the chameleon: its eyes can move independently of one another. This living fairy-tale creature was unknown to science until the 19th century, because it hid in the ancient jungles of Congo.

But I later came to realise that I didn't have to look so far afield to find adventure. Nor was it necessary to look among the mammals, even though they are the easiest to identify with. Another, larger group of animals had come first, and reading about them was sheer science fiction.

There were creatures that could have five thousand eyes, ears behind their knees, taste buds in their feet and a three-dimensional sense of smell. Their language might rely on chemistry or vibrations and still be truly sophisticated. Even 200 million years ago, these creatures had belonged to a high-ranking group of animals that went on to become extremely successful. Today, all the members of this group together would weigh three times more than the sum of all mammals, fish, reptiles and birds. The group also boasts more species than all other animals put together, and on an individual level there are a hundred million times more of them than there are humans.

In short, insects represent the standard size of life on Earth.

Insects had conquered the air long before any small dinosaurs tried out their new wings. Dragonflies first flew 300 million years ago, and the oldest known butterfly fossil is 250 million years old. Since insects are tiny and abundant, grow quickly and mate early, variations arise quickly; since they can manage with little food, they have survived Earth's catastrophes better than others. While large dinosaurs went extinct, bees, ants, beetles, grasshoppers and lice recovered relatively quickly. At the same time, other species became dependent on insects – not least birds, who evolved from dinosaurs, and the flowers that grew from seeds in the scorched earth. Eventually insects were so tightly woven into the fabric of Earth's environment that it would not survive without them.

Unfortunately, they're in trouble with us around. They seem so unlike us, and it's not easy to crouch down to the perspective of such tiny, fleeting lives. The ones that get really close to us, like mosquitos and fleas, we want nothing to do with. Thus, insects have become a world reserved for devoted experts. I'm not one of them, but I'm happy to listen to those experts who wish to spread their enthusiasm. And I've come to understand that it is ultimately thanks to the insects that springtime is filled with birdsong and flowers.

So far, the ground mostly seemed to be covered with old leaves and fallen branches from the springtime storms. To invite the green in, I should probably clean up the property a little. In the tool shed were abandoned implements for every season, from loppers to an ice auger; now I could take the opportunity to inventory these treasures as I looked for a rake.

But I also discovered something else. Next to a sledgehammer lay a few abandoned wasp nests. When I picked them up, they felt so light they might have been made of dust and minuscule wings, even though they were meant to hold a growing family. How could such weightless nests be so sturdy?

To take a closer look at their construction I brought them into the cottage, which, as it happened, had probably contributed to the nesting material. Perhaps it had been chewed from the door on the south side, where the paint was flaking. Then again, this could hardly be considered damage, for the amount of material taken was minimal and the result masterly. No wonder wasps were the world's first makers of paper. Here they had produced the thinnest kind I had ever seen, in the shape of a round lantern. I gently placed the nests on the kitchen table and removed the outer layer. Inside this globe hung a finely crafted plafond full of six-sided cells. Some were empty, but still resting in others were dead larvae. As an adult, each would have had recognisable individual features, and each would have been born with the skill of making paper that they would then fill with life of their own. Wasn't that a kind of poetry?

The half-grown wasps looked so young and innocent, lying there all bundled up in their cells. Could they help introduce my sister's grandchildren to all the birds and the bees and the flowers and the trees and so on? Although these young didn't personally have anything to do with the facts of life, there was a connection by way of their long family tree. About 140 million years ago, a few of their insect-eating foremothers got tired of chasing flying prey and decided to gather protein in the form of pollen instead. And this would prove to change them and the flowers, and the trees.

Rooted plants must deal with their erotic needs by messenger, and until that point the wind had carried pollen to female pistils. But the wind was fickle and high-flown, so a massive amount of pollen had to be released for any of it to reach its destination. Pollen-gathering insects made much better couriers. Since they sometimes had trouble finding the flowers under all the dinosaurs, the magnolias and water lilies helped by decking themselves out in petal skirts. Other flowers followed suit, and in so doing increased their attractiveness with an irresistible nectar.

Then something happened to the newly vegetarian wasp ancestors. Their upper lips and mandibles were reshaped into a straw that could more easily suck up the nectar, and thus they were transformed into bees. For the 130 million years since, flowers and bees have been trying to satisfy each other's needs; the flowers with their sweetness and the bees with their flight. That looked to me like love and, in any case, it created the garden of paradise in which we humans would one day arrive.

Perhaps the wasps' contributions to that garden were less evident, but without them there wouldn't have been any bees. Since they, too, like nectar, they sometimes help with pollination, and they feed their larvae insects that we consider pests. The venom in their sting even seems to be less potent than that of bees. So why didn't they ever become popular? Is it because they aren't as hairy?

Many things in life can hang by a hair, especially for bees, since their hair is necessary to capture pollen. In a delightful manner, it also helps in their interaction with flowers. As a bee flies, each forked strand of hair receives a positive electrical charge, while the flowers below have a weakly negative charge. So a small force field arises between the two, intensifying their encounter. They quite literally turn each other on.

The fact that hair preserves warmth was no advantage for honeybees in tropical climates, but it's a different situation for bumblebees. The temperature was dropping drastically in the Himalayas when they evolved about 40 million years ago, so a coat became a necessity. Thanks to that coat, bumblebees turned out so hardy that they can still be spotted near glaciers. And while honeybees prefer not to linger in temperatures below 16 degrees Celsius, bumblebees are ready to fly when it's just a few degrees above freezing. A bumblebee queen can even overwinter beneath the snow-covered ground, thanks to her fur and the glycerol in her blood that keeps it from freezing solid.

She likes to dig herself into a north-facing slope so she won't wake too early. By the time the spring sun has warmed that ground, a few flowers will have come up. The first breakfast of the year is traditionally taken in a couple of pussy willows, whose fuzzy blossoms seem akin to her own fur. The female flower provides energy-rich nectar and the male flower provides nutritious pollen, and that's just what she needs if the eggs she lays after last year's mating are to develop. But first she must find a safe home for them.

I had certainly noticed that the bumblebee queens were already waking up. After my encounter with the one who ended up in the rain gauge, I saw a number of bumblebees searching the property, surely on the hunt for a nesting site.

They were drawn to rather varied areas. The dream house of a large earth bumblebee is an empty mouse nest with some leftover grass insulation, and if she finds such a spot she's prepared to go to battle with any lingering mouse. A tree bumblebee, however, prefers higher locations, perhaps in the wall of an old building, with light insulation.

Sure enough. As I cleared a few dried stalks of lemon balm at the corner of the cottage I heard buzzing nearby. Then, silence. After a few minutes it returned again, and a bumblebee popped up at the lower edge of the wall. She was a little reddish, just like the one I'd rescued from the rain gauge, so it could in fact have been the same individual.

Might she remember me? Strangely enough, bumblebees can recognise humans. Or did the place seem vaguely familiar to her? Since it was a tree bumblebee, she could, in theory, have been born there the previous year. Now she crawled back in under the siding to stay for a while. This was just next to a bench where I was planning to sit down with my papers; during the warm half of the year I like to work outdoors. The light charges me like a solar cell as the insects buzz like a dynamo. If the bumblebee intended to make a home here at the corner of the cottage, we could keep quiet company together.

Whatever the case, apparently she was going to live right next door to my family, so knowing something about my new neighbour felt reassuring. Now I was doubly grateful to the researchers who had investigated the activities of the bumblebee. The enthusiastic ecologist Dave Goulson had even equipped them with tiny transmitters that registered their flight, and, like others, he had also paid careful attention to what went on in their nests. Thanks to this, I had some idea of what life in that wall would look like during the coming spring.

Unlike certain squirrels, a bumblebee doesn't need much space. A handful of insulation will do. The only furnishings in her nest are a collection of small pots she prepares out of wax from a gland on her belly. Once they've been shaped by her jaws and front legs, she fills them one by one with the harvest from various flowers. One little pot is heaped with nectar – good to have on the days she can't fly out. The others are stuffed with pollen and nectar kneaded into a dough, and on top of that she lays some of the many eggs she carries in her body. Once she's inspected and covered the pots, she lies on top of them like an incubating bird.

The fur on her abdomen is so thin that, like a bird's brooding patch, it allows close contact with the eggs. They need to be kept at 30 degrees Celsius, and she has a good sense of temperature. If her fur isn't enough, she can increase her body heat by vibrating her wing muscles. This raises her temperature when she flies, as well, so she is essentially warm-blooded.

After a few days of brooding the larvae emerge from the eggs, and once they've eaten their fill of the pollen stored in the wax pots, they spin cocoons where they will spend a few weeks transforming into pale bumblebees. As soon as they've wiggled out of the cocoon shell, they crawl to the nectar pot to gather energy, and then they move close to their warm bumblebee mother to dry their limp wings. The first bumblebee children are few in number and quite small, since they still have minimal resources, but the mother needs to gain helpers quickly. From now on, she will focus entirely on laying eggs, so within a few weeks there might be hundreds of young bumblebees.

I, too, had to plan to welcome a family, because even though we would be taking turns in the cottage, it had to have room for two generations. So the bunk beds were joined by a sofa bed assembled by a carpenter, and I myself dragged branches to the steep edges of the property to create low barriers. Hopefully this would deter the youngest from any spontaneous but perilous adventures.

The baby bumblebees in the wall, however, would be completely unprotected once they left the nest. While they were allowed to stay inside it for a few days, caring for new pupae and guarding the entrance, they would later have no choice but to leave and gather food. This is an enormous task for a young bumblebee. Outside meant great tits who had learned to scrape bumblebee stingers off on branches, and there were few flowers to visit in the spring. In dry weather there might also be a shortage of nectar, and finding it in the first place was a saga unto itself.

The first flight out begins with small circles around the nest area, for orientation purposes. All the distinctive features around it are carefully imprinted so the bumblebees can find their way home. Later, should something around the nest change, they are stumped. If, for example, you were to put a chair nearby they must perform another round of orientation flight to adjust their inner map, and if the chair is later removed they become confused all over again. I would really have to remember not to rearrange the furniture too much near that corner.

In general, bumblebees seem to notice everything. They prefer to collect pollen from different kinds of flowers so the larvae in their nest receive a balanced diet, and that means finding their way to and from those flowers. Thanks to fine details, they can cover a large amount of territory. The thousands of facets in their eyes see the world from slightly different angles, working together to convey information about distance, speed and route as the bumblebee flies. At the same time, roads, waterways and fields provide landmarks for orientation. All the while, the antennae are attuned to Earth's electromagnetic field and can react to minuscule changes in humidity, temperature and wind. What's more, they log every scent and can tell if it is coming from the left or the right. When it's time to make a precise landing on a flower, the antennae instead register the surface pattern of the petals.

How can we see bumblebees merely as peaceful *bon vivants*? They're efficient super-pilots with navigational instruments that don't even exist in modern aircraft. Thanks to these instruments, they can hold a straight course at 25 kilometres per hour even in a stiff crosswind. They're also the most industrious of all bees, making seven foraging trips per day and visiting four hundred flowers per trip. Since they fly even during the cool morning and evening hours, they often rack up eighteen hours of work in a day.

Behind this efficiency are proven methods. They remember the locations of half a dozen habitats and recall at which times the flowers give the most nectar. Then they schedule their visits accordingly and move from favourite spot to favourite spot in a rational manner. Should someone else recently have visited a flower, they bypass it the instant they sense the traces. Each time they land, they follow the same routine. The nectar is sucked up into a special receptacle in their body even as the grains of pollen caught in their fur is combed towards the pollen sacks on their back legs. It's important for the contents to be evenly balanced so they don't fly in circles, as the loads they carry can almost equal their own body weight.

Only when daylight begins to fade do they end their workday. Already when they left the nest in the morning, the three simple eyes on the tops of their heads had taken a reading of the position of the sun by gauging the intensity of its light, and the same happens again on the journey home. Then they know how much time has passed and which angle they should now take relative to the sun. A bumblebee has even reportedly found her way home from ten kilometres away, although her journey back took two days. At the same proportions, a human journey of that distance would be like taking a round trip to the moon.

So, where does it come from, this strange claim that bumblebees actually couldn't fly? Presumably it's a result of comparing them to dragonflies or gliding aircraft. But bumblebee wings move more like the rotor of a helicopter or like rowing oars. When the edge of the forewing angles up in flight, it creates an air spiral that provides lift. The disadvantage is that wing speed almost matches the revolution speeds of a racing motorcycle, so it's a technique that demands high amounts of energy. Some of the gathered nectar is usually consumed as fuel even during the journey, and this means they need a lot of nectar.

The flowers on the property made for excellent bumblebee resources, and they surely knew it. They love the blossoms of blueberry, lingonberry, heather, blackberry and raspberry; they love the flowers of old farm weeds and perennials, and herbs like wild mint and lemon balm – the very plant growing next to the tree bumblebees' nest. And dandelions. The bumblebees enjoyed resting in those yellow blossom baskets with their sun-warmed nectar, and I liked to sit down beside them to listen to the hum of their wings.

The bumblebee's entire body can be a musical instrument. Wing muscles quake like guitar strings, and with each impulse the wing flaps twenty times. All in all, bumblebee wings beat two hundred times per second. When the buzz of their wings blends with vibrations from their back plates and the membranes of their respiratory openings, it sounds like singing.

I noticed that it also took on a rhythm that described its very movement. The tones lowered as the bumblebee slowed down at a flower and paused briefly for nectar. Taking off required extra wingbeats, so the tone shifted again.

It was fascinating to find that the business of life could create sound. Every insect has a frequency, from the bumblebee's bass to the whining descant of a mosquito. The pitch is determined by the breathtaking wing speed. Wasp wings beat a hundred times per second; bees, two hundred; flies, three hundred; and mosquitoes, six hundred. Singer Gaby Stenberg made recordings of insect sounds and from them created a musical scale. A horsefly was C. A wasp – C sharp and D. A large bumblebee – D sharp and E. A bee – F. Another wasp – F sharp. A small bumblebee – G, G sharp and A. A flowerfly – B flat and B. A small bee – C. Altogether they formed a musical alphabet of wingbeats.

The bumblebees could hear this wing music better than I could, even though they didn't have ears. Again, they had hair to thank, or at least hair-like organs that captured even the tiniest vibration in the air. With their help, insects can perceive sounds at much higher frequencies than we do and can feel the air itself tremble with movement. Or, indeed, with longing. Female mosquitoes lure males with the tone of their wingbeats, so to be safe their wings are equipped with amplifiers. No wonder we can hear them on a summer night. At least to male mosquitos it is sweet music, and they immediately adjust their own wing sounds to the same frequency. For when a male is in tune with a female the two of them can mate.

The sound of bumblebee wings soon began to feel like home, but I also heard the buzz of wings in more unexpected places. One day, after coming across some red paint in the workshop, I decided to freshen up the flaking south-facing side of the cottage. Just as I was about to start painting, a pair of bees appeared before me. I backed up a bit to see where they'd come from, and they approached the door as if they wanted in. Strange. But after a while I realised we had different entrances after all. They had made their home in the casing of the door.

Aha, so multiple kinds of bees lived in the south wall. Was I on the verge of painting some sort of beehive? I had seen photographs of Slovenian beehives that had been painted like old farmhouse cupboards. The images were colourful in order to be visible through a swarm, and in a robust style often portraying beekeeping or a biblical garden of paradise. Such paintings could simultaneously demonstrate ownership and brag of a wealth of bees. But after all this was no beehive, for the inhabitants of the door casing were wild, solitary bees. They were called red mason bees, and, like the tree bumblebee, had a slightly reddish coat. Meaning all of them matched the colour of the house.

The red mason bees must have been there since the summer before, when a lone female made her way into the wall to lay eggs in peace, just like the tree bumblebee. But from that point on she fulfilled her motherly duties in a different way. She left all her eggs in separate, small cells, each with its own stout supply of pollen, mostly from maples and oaks. Then she closed up the nursery and flew off. Even if she herself didn't survive the winter, her children would make it through in the warm south-facing wall, and obviously they had. They'd waited like tiny commas as the months passed – and the tradesmen went back and forth through the door – slowly developing with the help of last year's tree pollen. To me it seemed they gifted the house with a life of their own.

They're said to be a peaceful, even child-friendly species of bee, and solitary bees in general do tend to be less aggressive than those who have a common nest to protect. Perhaps some of the door-casing bees had already emerged when the pussy willow began to blossom, for they tend to coordinate in some mysterious way. The males would have been closest to the entrance so they could leave quickly. If there were to be new bees for next year, they had to mate as soon as possible. Still, they weren't abrupt about it, because solitary bees can be highly attuned to the needs of others. The males would gently stroke the females' antennae until they received a spirited 'yes', and afterwards they rested together for a long time. Later, perhaps, the door casing would become a nursery again, although no one in the house would notice the growing children before they flew the nest in the spring.

In the bumblebee nest, however, a summer of busy family life awaited, right next door to the cottage's combined bedroom and living room. To be sure, the bees would probably be quiet neighbours. Good-natured bumblebees are beloved by poets and children's book authors alike, and their lives are characterised by a feminine sort of care. Brutal alpha males are largely absent from the insect world, where egg-heavy females are larger as a rule, and bumblebee nests are matriarchal to boot. They aren't led by an old matriarch as elephants are, however, because a bumblebee mother is too busy laying more eggs. Instead, the daughters must take care of themselves and their younger siblings, and will do this in harmonious sisterhood until the end of summer. But the drama that then awaits them would make suitable subject matter for a Greek tragedian.

In high summer, all would still be idyllic. Hot days are spent napping together, and if the temperature in the nest rises above 30 degrees Celsius a few individuals will post up near the entrance to fan in air with their wings. In the centre of them all, the bumblebee mother sits on her new eggs, fed by her sweet daughters.

All along, time is slowly making changes to everyone's hormones. The mother is aging, and the future demands more youthful forces. To comply with this urge, the daughters begin to give special care to a few eggs that will become new queens. But they must also be fertilised by a male.

Up to this point, all the eggs have been fertilised by the sperm the queen stored after last year's mating. With a chromosome from a drone, they become females, but without they become males. And now the queen begins to lay unfertilised eggs with a single chromosome. It's immediately obvious that the bumblebees emerging from these eggs are different. The pattern of hair on their faces looks like a beard and mutton chops, and it's also clear what they're after. Outside the nest, they surround themselves with seductive scents, preferably that of lime blossoms, which can make bumblebees rather tipsy. They leave their delicious scents in long trails on bushes and trees.

Sure enough, the newly born queens quickly fall for the lecherous males. In the considerate way of the bumblebee, they make sure not to sting their partner during the act of mating, and when their connected bodies plunge to the ground the queens are filled with new life. If this life is to have any chance of survival, they must store energy of their own before hibernation, so the queens never return to their siblings in the nest.

Life is changing there as well. The flowers are giving less and less nectar, and there's more and more tension between bumblebee mother and daughters. When the daughters realise she has laid unfertilised eggs, they begin to do the same, for the whole nest is under the influence of hormones. But their egg-laying makes the once benevolent bumblebee mother angry. Perhaps it's because this little stunt goes against the principles of their society, or perhaps it's because she is genetically closer to her sons than second-generation offspring. In any case, she bites her egg-laying daughters and eats up their eggs. It doesn't matter that those are her grandchildren. The daughters respond by ganging up on her and eating up her male eggs. It makes no difference that those are their brothers.

Even though all of this is determined by biological patterns of action, it's still a sad story in the end. By the time the ever balder and more fragile bumblebee mother tries to retreat from the skirmish, it's too late. As in a classical tragedy, she is either killed and flayed by her daughters, or left to starve to death in her disintegrating nest.

The only survivors are the fertilised young queens that have already left the nest. Certainly, some of them will also eventually starve or be eaten, if they don't moulder in some damp hibernation spot. But some of them may yet wake up with the first flowers in spring.

What drama can arise within a family. At least the solitary bees of the door casing avoided such a life. They had, on the whole, fewer opportunities for conflict, since their young had to fend for themselves. With no new generation or common nest to defend, they had never developed social behaviours.

Yet, like bumblebees, solitary bees are actually better pollinators than honeybees, and since they have only themselves to rely on they can be quite clever. For instance, a solitary bee has been observed removing a nail from its intended nest all on its own. When solitary bees are brought together, they can also help each other out with difficult tasks. But if they are incorporated into a colony with divisions of labour, it seems that their broad repertoire shrinks, rather like when craftspeople become factory workers.

So, life can function well in different constellations even among bees, for single life and family life each have their own set of advantages. But it's the reclusive traits that are dominant. Out of the nearly three hundred species of bees found in Sweden, all are solitary except for the honeybee and about forty bumblebee species. And even bumblebee societies begin with a single queen, although her daughters later remain in the nest. Why is that? Why don't they build their own small families, or choose an independent life as solitary bees?

The childlessness of the bumblebee daughters isn't that unusual, for most animals die without leaving behind any offspring. No, what's worth considering is that helpful daughters at home can form the base upon which a society is built. In all their simplicity, they show that life isn't necessarily ruled from above. These bees find that quite the opposite is true, and everything depends upon a humble sisterhood that might just resemble love.

Cooperation is at its most evident among the honeybees. Aristotle considered their system a positive example, although he preferred not to view it as a matriarchy – after all, bees were armed thanks to their stingers. Only under the microscopes of the 17th century did it become clear that the bees' kings were really queens, and the idle drones were males. Still, the queen didn't rule, so what was the source of the order that gathered them all into a nearly organic community? Was it the number of individuals, or something more mysterious?

In some remarkable way, a beehive unites two vastly different worlds. One is built upon community. By brushing against the queen, all bees receive substances that stimulate their instincts for care and building. Their bodies quite simply adapt to the various needs of the hive. The very youngest care for their larval-stage siblings, so they excrete a protein-rich glandular secretion the larvae can feed on. Their next task is to prepare the nectar other bees bring home, and at that time their bodies produce an enzyme needed for such a process. After a few more weeks, they will fly out to gather food, so they generate hormones that turn them into foraging bees. Their bodies already contain everything needed to help their thousands of sisters. The whole structure of the hive is tailor-made for the community, and all have developed in the same honeycombs.

With its six-sided walls, the honeycomb is a geometric wonder. The hexagon is the shape chemists use to demonstrate how the molecules of life are constructed by atoms, and how those very molecule figures go on to be combined into larger patterns. The six sides also have a practical advantage in a honeycomb. Since each cell shares its walls with neighbours, this structure uses the minimal amount of material and its weight is spread out evenly. The bees that build honeycombs seem to understand this. They don't reinforce only the weak sections; they strengthen other parts at the same time. The cells must be sturdy, because the larvae that will develop inside them can grow to a thousand times their original weight.

So how did bees arrive at the hexagonal method of construction? The most remarkable fact is that the six walls arise on their own as the bees work cooperatively. Unlike the wax pots of bumblebees, the cells are not built one at a time, but by a number of bees moulding their wax all at once. They work so close together that the wax around each bee melts with the heat and is united with that of the neighbouring worker. Because they keep a certain distance from one another, they end up with six walls of equal size. The honeycomb perfectly illustrates the cooperation within a hive.

But the cells inside will be filled with more than just new siblings. They must also store something that is gathered outside the hive, and in order to search for it each bee must fly out on its own, into a world that is wildly different from the regimented environment inside the hive. Where the hive is crowded, dark and systematic, the world outside is boundless, bright and ever-changing. How can they possibly find their way?

Like bumblebees, honeybees can recognise their surroundings as long as they're within about a hundred metres of the hive. Beyond that, they must navigate at an intersection of time and space. They tell time by the sun, following a kind of sundial where time is connected to different tours among flowers. Even back in the hive they will have studied it by way of a narrow beam of polarised light, and from this they learn to differentiate between six points of time in one day. More than that is *always*. But they also understand the time when no nectar or pollen can be collected, which can lead to starvation – the *nothing* that can pose a threat to all life in the hive.

Then their sense of time must be combined with the landmarks of their territory and the clues from their senses, and that takes flexibility. Wind and weather change constantly, and vegetation can change by the week. At the same time, they maintain a running dialogue with their sisters in the hive, because if their common resources run short they must turn to flowers that offer less pollen. In the midst of constant development, they must recognise various flower species, know which are giving the most pollen at the moment, and track them down. In other words, they must both learn from experience and make their own decisions.

*   *   *

The foragers' harvest is collected inside the hive. There, nectar is prepared and pollen is sorted into different colours. There, too, the bees prepare a putty that will seal any cracks in the honeycombs and protect the nest. The materials are largely gathered from the resin-coated buds of broadleaved trees and the pitch of conifers, and it seems to be some sort of life essence. Among the hundreds of ingredients there are even traces of silver and gold. This putty can kill not only viruses and bacteria but also fungi, so if any intruder dies inside the hive it too is covered in the disinfectant material. This miraculous, hive-protecting substance is called propolis, which in Latin means 'for the community'.

But the most important product of the hive is found elsewhere. It's a sweet, golden substance that is collected in the wax cells and serves as insurance for the future. Honey. It runs a splendid gamut from white to amber to bronze, as the different flowers it's made from determine its colour and aroma. Early summer honey is pale; in autumn it's darker. Clover blossoms produce a mild flavour, lime blossoms a fresher one, and heather a more aromatic one.

Still, no flower can entirely describe honey, for like propolis it has hundreds of ingredients. In addition to vitamins, minerals, antioxidants, lactic bacteria, amino acids and formic acid, there are special enzymes added by bees. The nectar is brought into the hive by passing from mouth to mouth, so one sugar molecule touches any number of bees before it reaches the honeycomb. Thus, honey cannot be understood to result from any individual bee or flower. It's born through an interplay of different species, individuals and times.

Is that why honey has always seemed to have a particular glow? It made its way into cave paintings and Babylonian texts. It was there in the Old Testament promise of a land of milk and honey, as well as in the Quranic vision of paradise. In Egypt, where bees were said to be tears of the sun god, honey was thought to bring long life. And it certainly was long-lived itself. Some honey found in a three-thousand-year-old Egyptian grave apparently remains edible.

The wax surrounding the honey seeped into human culture as well. It brought crayons to art and moulds to sculpture. Roman writing tablets used a beeswax layer; when warmed it could be made smooth again, ready for new words. Beeswax sealed ships against seawater and made clothing watertight. It also made the earplugs that saved Odysseus' shipmates from the lure of siren song, and it anchored Icarus' wings, although he forgot that wax melts if you get too close to the sun. The dark Earth, however, has been illuminated by beeswax candles for millennia.

The Berber language is said to have a special word for something that's born when the sun touches a honeycomb. It's like a symbolic image, but for what? Is it the essence of a thousand flowers and the many paths that lead to them?

When I pictured a honeycomb, it reminded me of the

faceted eyes of bees, since those too have six-sided parts. Each section provides an angle of its own, and working in conjunction they help the bee find its way to a flower. The brain's vision centre is made up of similar parts, although in that case they are nerve cells that allow the bee to see and react. Nine hundred thousand such nerve cells fit inside a tiny bee brain, all reaching out for one another, for even among cells it takes a joint effort to capture the signals of the world.

On every level, honey depends on cooperation. In three weeks, a single bee can only gather, at most, a quarter teaspoon of honey, and is then worn out. It takes two million flower-visits to fill a jar of honey.

So why all this labour? Like the ancient Greek gods, bees really like nectar. They can even get drunk on it; when fermented, its alcohol content can reach 10 per cent. Intoxicated bees have even been studied in the hope of finding clues about alcohol abuse. What scientists discovered was that alcohol made bees bolder. Carl Linnaeus's brother Samuel gave his bees wine mixed with honey to help them better fight off robber bees who were stealing their stash of pollen and nectar. On the other hand, sometimes tipsy bees can't find their way back to the hive, and even if they did they would be turned away by guard bees who don't hesitate to snip off the feet of notorious drunkards. The nectar isn't meant to be an intoxicant. It is only intended to provide energy for gathering even more nectar and pollen. Eventually, it will all be transformed into something that spans time and space, and doing so requires cooperation, which in turn demands something more: communication.

Aristotle observed that bees dance, but it took another two thousand years before it was understood that the dance itself has meaning. In the mid-20th century, Karl von Frisch interpreted the dance and concluded that it was a complex language.

Von Frisch grew up in a Vienna buzzing with intellectual life. He came from a family full of professors who considered big questions to be natural, and after forming a string quartet with his brothers he also came to realise the importance of interplay.

But he wanted to communicate in multiple ways. He had a parrot for a companion. It liked to sit on his shoulder, bite at his pens and sleep next to his bed, and the first thing von Frisch did each morning was try to speak with his bird. He kept about a hundred other animals besides the parrot, so it seemed natural for him to study zoology.

He began by studying fish but gradually pivoted to bees. In both cases, his first discoveries had to do with the creatures' senses. In the case of fish, the senses of taste and hearing seemed to be most important, while smell and colour were important for honeybees. They use scents to find their way to flowers, which they recognise once they're close by their colours and shapes. A special attention is paid to the difference between firm and vague contours.

Von Frisch's research team also noticed that bees have a vivid relationship with time. If they were regularly given sugar water at a certain hour, they would punctually show up at the very spot where they had been served. And it was clear that they had their own form of communication. Von Frisch was on the case.

But while his research advanced, human society was headed for one of its most major breakdowns. Nazism was spreading in Germany, heading for a world war and the annihilation of a people. Von Frisch was a professor at the University of Munich at the time, and he was so absorbed by his research that the events of the world outside mostly seemed like dull thunder in the background – until a swastika-stamped letter arrived at his institution. When he opened it, all he found inside was a brief message. He was a quarter Jewish, so he would be dismissed from his professorship.

It was a double blow. If he was forced out of the university, his research would be cut off just as it was about to reach a breakthrough. Cooperation was as important within his research group as it was among the bees, and several renowned colleagues pleaded for him to be allowed to stay, but it was all in vain.

Instead, the bees of Germany came to his rescue. They were being ravaged by hordes of the intestinal parasitic fungus *nosema*, which literally ate its way into the bees' bodies. Twenty-five billion bees had already died.

Bees are a key species in ecology, and a large percentage of the human diet depends upon their pollination. Food shortages were already a problem in 1940s Germany due to the war, and now the situation grew more serious. At the same time, rumours were swirling that the Soviet Union was beginning to train bees to greater speed and efficiency. Shouldn't the Reich be able to do the same? After all, bees were a shining example of citizens willing to make sacrifices for their society. From the Reich's perspective, von Frisch's research on bees' language was uninteresting, but applied research that might lead to replenished food stores was urgent. Could von Frisch solve the parasite problem? His removal was postponed in the hope that he could, and in the meantime he was allowed to continue his other research.

It was a paradoxical situation. While decoding experts tried to figure out the enemy's messages, von Frisch was searching for what was, in the best way, an unhostile inhuman language.

It takes time to learn new languages, not least when they're in no way similar to the one you've already mastered. But gradually von Frisch and his colleagues managed to penetrate the communication of the hive.

First and foremost, it's about flowers. The scents that tell of them are easily borne aloft, and they can also be understood by other species. The flower language of bees, however, is a sophisticated code that shares some traits with art.

It describes both the quality and quantity of the flowers' nectar and pollen, as well as the path to reach them.

All of this is danced out on the honeycombs in the hive, forming a map of symbols. Simple circles indicate nearby flowers; infinity signs mean flowers further away. The length of the dance denotes the distance to them or the time and energy the flight will demand. For instance, if the bees must fly into the wind, it will take more energy. The direction is shown by a centre line within the circle. If the messenger bee runs up the line, they should fly towards the sun, and if the bee runs down the course will be the opposite. If the dance moves diagonally and to the right, the direction of flight, likewise, should be to the right of the sun, at the same angle as the dance takes in relation to the vertical line.

Even the bee's bobbing hind end gives information about the direction, travel time and flowers. The more zealously the bee sways, the better the nectar and pollen. At the same time, the flowers are compared to the hive's existing resources. If food stores are slim, the dance includes paths to less bounteous flowers. Even though it's all about distance, the dance is filigreed with detail.

It's about riches that exist in light but is itself performed in darkness, so it must be understood through vibrations. The bee's wing muscles tremble as it moves on the honeycomb; the wings are able to buzz at the same frequency as flight even when they're folded. This provides additional information about the flight and at the same time creates yet another language. It's like a Morse code consisting of vibrations with bursts of tone at a rate of thirty times per second, as well as pauses and shifting pitches. A dancing bee thus braids several languages together.

These languages are not just for describing a path to flowers. Should the hive get too warm, for instance, it needs to be cooled, so the bees must bring drops of water rather than nectar and pollen. When other bees move their wings above the water, they produce a sort of air conditioning. If they need a refill, the way to a water source can be described through dance.

The dance can even convey information about an entire environment. This is especially likely to occur during a swarm, when the old queen forms a new colony along with half the bees in the hive. Since bees are systematic creatures, scouts are sent out beforehand to have a look around the neighbourhood. Their reports back must include various factors. How big is the potential new home? Is it dry and free of other insects? Are there old honeycombs from past bee colonies? What does the entrance hole look like? How far is it to flowers and water? All of this must be described through the dance.

The volume of the cavity is so important that the scout bee might spend forty minutes methodically inspecting the walls. She continually checks the changing relationship between angles and can arrive at a mental cross section by remembering the distance between them. The position of the entrance hole is measured in the same way. It's also crucial to have a water source nearby, but if it means flying across a lake the suggested location will be rejected by bees with local knowledge who can assess the situation.

Thus, at swarming time, the dance describes something entirely different from flowers. What's more, the description of the location and its surroundings must be exact, for many bees will have been out scouting and all their suggestions must be compared before one can be chosen. Still, it's not exactly a competition, since all the bees allow themselves to be influenced and solicited by one another. In the end, the ones with the most support bring the whole swarm to the recommended spot.

Bees, then, have had a functioning democracy since before the Greeks even created the word for it. No one and everyone makes the decisions in a bee society – a far cry from the king that was once imagined to rule a hive. Bee communication is, in fact, a type of dialogue, since the scouts need an active audience, and only urgent questions are discussed. When the hive has everything it needs, there is no dancing. It isn't meant as entertainment; it's a tool to deal with the conditions of life.

Dances have arisen just about everywhere on Earth, with different purposes. They can be part of mating rituals or religious rites; they can create community or become an art form. Something of all this is evident in the dancing of bees. As in mating rituals there is a fertility aspect, and as in religious rites there is an enigmatic, unifying essence. There are definite steps to the choreography, but, as in folk dances and dialects, there can be local variations.

Above all, dancing is a well-refined language that suits the bees' ability to interpret movements using every sense.

While to me an insect's flapping wings look like one big blur, the bees can see them moving up and down, even if they're moving two hundred times per second. If you were to make a film for bees, twenty-four frames per second wouldn't cut it. It would take ten times that rate to give them a sense of smooth motion.

In one sense, bees are realists. Their senses are so attuned to nature that abstract shapes like triangles and squares mean nothing to them, so they don't always notice such things. True, researchers once managed to teach them telling paintings by Picasso and Monet apart, just as had been done with pigeons. But it's likely that what the bees saw were the differences between strong and weak lines, because that's the sort of thing they typically notice in flowers.

In addition, the shapes of flowers and their scents are linked, since both of these traits are sensed through the antennae. When other bees brush past a dancing bee in the darkness of the hive, they also notice the scent of the flowers the dance is about. And that's not all. The description of the flowers, and the path to them, affects them deeply. Round objects leave different impressions than angular ones, for soft and firm shapes have differing scent signatures. As a result, a landscape of shifting forms comes alive during the dance, giving shape to a three-dimensional world of the senses.

All in all, this is a fragrant and mathematical language, a marriage of poetry and land surveying. In mathematics all is clean and scaled back, abstract and exact, while in poetry so much is built upon sense associations and implied words. The unspoken gives rise to tension between the words, vibrating as in the relationship between the flowers and the bees. The bees' dance is a complete and natural description that reaches from within the flowers to the winds and cardinal points of the landscape. All of this can be conveyed to others in a map that is both poetic and precise.

Above all, this is the story of a common world. If a single bee wishes to ask for help or give encouragement, she can do so straightaway, using particular scents or pheromones. Such immediate appeals have surely often been the beginnings of a language. So where does it go from there? Does language develop in a certain number of individuals, or in a common home? Or is language created by way of a shared task that involves something beyond the present moment, such as gathering honey? Maybe it's all of this. In any case, bees have shown us one way in which a magnificent and unique language can arise.

The discovery of the language of bees was a sensation, and in 1973 von Frisch was awarded the Nobel Prize. He shared it with Konrad Lorenz and Nikolaas Tinbergen, who had also studied animal behaviour. Biology had long been concerned with the determination of species, but now bees were no longer limited to insect collections. Their dance became part of linguistic theory as well.

It was revolutionary to find that insects were capable of complex communication, but it also brought up troubling questions. After all, we humans had long considered ourselves superior to other creatures thanks to our use of language. Did this mean that we weren't the only higher-order species?

While von Frisch was researching the language of bees, the issue of advanced species had been thoroughly besmirched. Nazism's division of humans into higher and lower races was contrary to both morality and reason, and had profound consequences. On the other hand, it was unsettling to discover that other species could be highly developed, however advanced their language might be. And so the issue was set aside.

In fact, though, this was when the situation truly began to deteriorate for bees. They were used to moving among wildflowers in a small-scale world of fields and ditches, but in the post-war period small family farms gave way to industrial agriculture with innumerable varieties of pesticide. Often, this brought unintended consequences – it might turn out, for example, that a fungicide multiplied the effects of an insecticide many times over. While pests quickly developed resistance to such treatments, the birds that kept their numbers down were poisoned. Bees were similarly affected. Even after DDT was banned from the market it lingered in the ground and could be found in the pollen gathered by bees.

People understood the value of bees, of course, so bees too began to be industrially farmed. These days, factories produce millions of bee colonies, and even bumblebees have become a commodity. They are needed to vibrate the pollen loose in greenhouses full of tomatoes and berries, so the bees are packed up in boxes and loaded on lorries for jostling journeys that can traverse continents. The enormous fields of monoculture provide a uniform diet seasoned with ever-novel pesticides, so perhaps it's no wonder that the bees' resistance has taken a nosedive. The intestinal parasite *nosema* has been joined by the horrific *varroa* mite, which is also spreading to wild bees, and the honeybees themselves are starting to have trouble finding their way back from foraging in the field. Even worse, the number of bees and bumblebees in Europe has fallen by 75 per cent, and 90 per cent of the bumblebees in the United States have vanished. If pollinating bees disappear we will face disaster.

Is it the fault of pesticides, monoculture or the toll of transport? Is the bees' sense of orientation disturbed by cellular service towers, and is climate change a factor? Perhaps it's a combination of all such things, for large-scale thinking doesn't seem to be a good match for small-scale transmission of life.

Since there is less pesticide application in urban environments, people in some locations have installed beehives on roofs – indeed, even on cathedrals such as Notre-Dame. But above all, bumblebees and wild bees have fled to private lands where they now number greater than in the shrinking pasturelands. On small forested lots they can also avoid being forced to mate on an assembly line under the fluorescent lights of a factory, and can select their blossoming friends on their own terms.

* * *

Soon enough, the solitary bees in the door disappeared, which meant that they must have gone their separate ways after their romp in the spring sunshine. If they hadn't developed a language of their own, like the honeybees, it must have been because they didn't often need to share messages with others. Still, like all other bees, they naturally had their own inner world of flowers, foraging and memories. They seemed to have a good sense of the world around them; they always seemed to have a destination in mind when they flew out of the doorway casing. And they were excellent pollinators, even if they didn't tell each other that the dandelions were blooming to the east. It was enough for them to have individual knowledge of the best flower locations.

In the case of the tree bumblebees, the first brood emerged in May. Since they hadn't had much food to grow on, all of them seemed heart-wrenchingly tiny, but what they lacked in size they made up for in energy. A grassy path a metre wide led alongside the house and to the door, and as I was about to mow it with an electric lawnmower I'd inherited, an indignant little gang of guards came pelting out of the wall. I suppose the mower made their nest vibrate alarmingly, and apparently their territory extended all ways out.

Surely they'd had time to form their inner maps of that territory. Would I have recognised myself on them? After all, our senses were different. While we appreciated the same flowers, I couldn't enter their landscape of faint or strong scents, which sometimes combined in harmony.

Since scents were part of their way of communication, they also deposited their own odours on the flowers they lived with.

Our colour scales, too, were a bit different. They couldn't see red, so their world was slightly tinted with blue, but then again, they could see ultraviolet, and in their eyes it formed glowing nectar patterns and lent daisies a bluish-green shimmer.

In general, we seemed to have a similar world view, even though they were a thousand times smaller than me. The fact that they could even recognise people sparked my admiration, because bumblebee researchers must attach tiny numbers to their subjects in order to tell individuals apart. Had I now been incorporated into a bumblebee brain? It was no larger than a grain of salt and yet could hold a map a kilometre wide, created by hundreds of thousands of nerve cells that registered the most minuscule variations in scent, sound and light. This was the very union of something close and something broad that poetry can attain. Perhaps I was actually searching for the perspective of the bumblebees? Due to a tiny scar in my right eye, I do in fact have the impression of a pair of insect wings constantly hovering in front of me, as if there is a tiny pilot there. The thought appealed to me.

There's a hint of emotion in the bumblebee's sensitivity, and it would come to be confirmed in experiments. It turns out that trapped bees can feel fear. If they're not released, chemical substances can accumulate in their blood, causing them to die of panic. When, in other experiments, they were exposed to substantial shaking, they became apathetic. I could sincerely understand why being transported in crowded boxes on lorries went against their very nature. The researchers found that everything that happened to them could provoke agitation.

And eventually, half a century after the discovery of the honeybees' advanced language, humans would have to admit that bees, too, have consciousness. In 2012, leading experts in neuroscience and cognitive research signed, with great fanfare, what they called the 'Cambridge Declaration on Consciousness'. It was formulated in a scientific, matter-of-fact manner: 'Convergent evidence indicates that non-human animals have the neuroanatomical, neurochemical, and neurophysiological substrates of conscious states along with the capacity to exhibit intentional behaviours.' We are, then, not alone in possessing everything it takes to have a consciousness. It is also found in other beings.

This was actually a greater revelation than the discovery of bee language. Consciousness had long been thought unique to humans, but here that assumption was roundly rejected. In the hunt for the physical source of consciousness, researchers travelled far back in time, all the way to the common ancestor of arthropods. She lived 540 million years ago and was probably the first conscious organism on the planet.

Naturally, a scientific explanation was offered at the same time. Each central nervous system must involve a brain of sorts, and in both vertebrates and insects it took the form of an enlarged ganglion. This organ processed and coordinated sensations, which allowed the creature to orient itself and learn from its experiences. Put simply, subjective experiences were crucial to surviving all the twists and turns of life. Any animal with a nervous system, then, could in theory experience things like fear, anger, security and intimacy.

I took my coffee cup from the bench, where a few buzzing little neighbours had just flown by. Perhaps they were on their sixth trip and their two thousandth flower. I would never know what went on in their millimetre-sized heads, but I knew better than to underestimate them. When tested, bumblebees proved themselves able to solve novel problems, such as how to remove a lid from a source of nectar, and they could learn quickly from one another. If they were rewarded with nectar they could even perform feats far beyond what was necessary in their usual lives; rolling large balls towards particular goals, for instance. I had read that one researcher, exposing bumblebees to a psychological test, ended up assigning them the same IQ as a gifted five-year-old. At first I nodded along in agreement, but then I began to wonder how well versed he truly was in their lives. IQ tests on animals are still performed in laboratories, far from their natural environments. So was the researcher even aware of bumblebees' precise navigation and the way they made plans to visit different kinds of flowers when the nectar was most abundant? Did he understand that they could organise an entire little community; did he appreciate their general ability to survive a demanding life? If so, the five-year-old he was comparing them to must be a genius. And I wondered if any human could pass an IQ test created by bumblebees.

That evening when I went inside, I heard a humming in the south wall of the room. Was the bumblebees' winged heating and air conditioning system in use, or was something else going on? Although they don't dance in their nests, they do communicate. If they want to alert each other to any special flowers, they bring home a sample of nectar, and when the nest's resources are fading they can use pheromones, urgent nudges and buzzing to prompt others to follow them to a few fruitful locations.

In any case, something intense was happening in the wall. Didn't they ever rest? Their buzzing conversation lasted all evening, like a hovering tone right next to me. I would have loved to be included in their conversation, because I was fascinated by everything it might cover. During the summer the topics of birth and death, idyll and disaster would follow on each other's heels like a miniature epic. And even though the nest was much smaller than a beehive, it still contained honey. It was thin as juice and difficult to store; it only needed to last for the summer. After that, of course, almost the entire bumblebee family would disappear. But still – inside my wall rested tiny pots of honey. Although they weren't for me, they felt like a hidden treasure.

While the bumblebees were occupied in their flourishing pantry, I had an evening sandwich in the kitchen. There was, in fact, a jar of fluid honey in the cabinet. My sister, the gardening expert, had advised me to smear it on the damaged surface of a broken branch, and why not? After all, it was one of the world's most nourishing substances, full of ingredients from the plant world.

As I dribbled a little honey on a piece of crispbread, it gathered in the many little holes. I was reminded of honeycombs, maybe because they were lingering in my thoughts. Something about the six-sided cells was still occupying my mind, and the associations extended to other walls as well. Even this very cottage had six sides, although they made the shape of a die.

When I was a child, evenings were often spent with board games. I was only moderately interested in the games, but since it was cosy to gather around them I was still happy to play. In fact, a small matriarchy sat around the table, in the form of my mother, her sister, my sister and me. The others had inherited my grandfather's nearly mathematical love of games, and while they were absorbed in their moves I often let my thoughts wander. The die, for instance, was interesting. Although its six sides presented different values, each one could move the game forward. If I rolled a one or a two, I thought about the meaning in small things. With patience, even small steps could get you where you were going. When I rolled a three or a four, I thought about the mean. Perhaps there wasn't much exciting there, but means form the framework of statistics and indicate something universal. A five meant many steps forward, and if I rolled a six, which let my piece fly ahead on the board and gave me a bonus roll, I thought of a tailwind. It was something you couldn't control, and it could change suddenly. It was while playing these games, then, that I pondered life.

People have found five-thousand-year-old dice made of animal bones, so surely many before me have been fascinated by their construction. Each side has a unique value, and yet all of them complement one another. Add the sides opposite one another and you'll always arrive at the same sum. One and six are seven; four and three are seven; five and two are seven. Seven is an odd number that recurs in many places. Man sailed the seven seas and the world had seven wonders; the rainbow has seven colours and the week, seven days. The virtues and deadly sins number seven. In nature, it seems that many birds are able to count to seven, so perhaps it's a figure the brain can easily comprehend.

For honeybees, a die would probably have more visual associations. The dot of a one would remind them of the entrance to the hive, and the symmetrical rows of a six were not unlike the hexagonal cells of the honeycomb. But bees are also able to count to six – that's how many times of the day they keep track of. More than that turned into *always*, which to them might encompass many different lives. The starting point was a lone queen, and from her a society grew. For bumblebees, life lasted over a summer, while in a beehive, thanks to the honey they'd gathered, life could go on even as the world became cold and barren.

Interplay can begin with being united around something seemingly simple. A die can suffice. In my family, our board games would later be supplemented with more artistic glass bead games, which also provided a gathering point. Music, dance and poetry too have traits of a uniting language. It seemed to me that some of this could also be found among the bees.

I took my honey crispbread out into the evening light, where life was humming with insect wings all up and down the scale. Those tones could create melodies I was unable to hear since they came from different spots and from half-invisible bodies. But they were there. All I had to do was look and adapt to a different scale.

Chapter Three

The Ants on the Wall

IT WASN'T HARD TO UNDERSTAND why bees were associated with a sun god. They move about in the light and nourish themselves with nectar, and their wings create music when they fly. They have a language of dancing symbols that links the insides of flowers with cardinal directions, and out of all this they create a substance from millions of condensed seconds of light. Everything about their lives is pure poetry. Like poets, they can be solitary or live in broader contexts, and they clearly demonstrate that each alternative has its advantages.

Their more distant relatives, however, have settled upon an extremely earthbound social life, no part of which suggests frivolity. They have no wings, no colours, no conversations with flowers, and no hair to which pollen can stick. They seldom go out alone for lengthy adventures, and instead of flying or dancing they march. Could I come to love them too? In any case, I had done my best to understand them.

I had, of course, noticed the ants during spring. When they awoke after their winter hibernation they were in great need of fortifying, and apparently birch sap wouldn't cut it. From the opposite side of the cottage walls they managed to smell their way to the pantry, where they crawled down into an open carton of juice. I very nearly swallowed them. Later I found ants in the sugar, even though it was stored in a metal container. They really wanted to insert themselves everywhere.

To rid myself of the caravan crossing the kitchen, I poured sugar into a bowl and set it far from the cottage. I did not wish to share the kitchen with ants, even if they were cleanly. The fact that the bowl later disappeared under mysterious circumstances is another story, and cannot be directly blamed on the ants.

Why didn't I have the same warm feelings for ants as I did for bumblebees? They're both descended from the same insect-eating wasp, although that ancestor gave rise to offspring with very different life paths. While the bees have tested out various constellations of life, ants have kept strictly to a communal life on the ground, with some rather ascetic traits. Where bumblebees move like squirrels in the trees, ants live like naked mole-rats in underground colonies.

Living underground provides excellent protection. But the most prominent explanation for the success of ants is their large number and their cohesion. The more ants there are, the more successful they become. Today there are fourteen thousand known species, and surely just as many still unknown. Each species has its own style of nest, allowing them to adapt to various environments, which in turn has allowed ants to establish themselves in every temperate corner of the Earth. Altogether, there are more ants than there are seconds that have passed since the Big Bang.

While bees are under threat, there doesn't seem to be any danger for ants. On the internet, I had mostly found them under the heading 'pest control'. Gentler sites advised me that they dislike cinnamon, pepper, garlic and bicarbonate of soda; sprinkling these across their path will stop them just like a barrier. Perhaps I should try that in my kitchen. A more radical solution was to put out pest-control traps, which contain a yummy poison that the altruistic ants would naively carry to their queen. When she ate it she would die, and without her the entire ant society would automatically collapse. I made note of this in case the ants became too troublesome.

There was something irritatingly familiar about their expansion. Was it that they seemed to mirror our own spread and increasingly urban lifestyle? Whereas bees live in a rural, small-scale sort of way, the ants' settlements seem like big cities. In relation to their size, their cities can even be larger than London and New York. I couldn't see the extent of their underground constructions on the property, but I must have had tens of thousands of lives under my feet.

Their structured, earthbound existence is full of energy, so I knew there must be ants dashing this way and that in a multi-level nest packed with storage rooms, paths, warehouses and dormitories. The nursery of eggs and pupae must be on the top floor, since they needed more warmth. But no ants like the cold, so after their winter torpor they typically shuffle one by one into the spring sunshine to thaw their frozen joints. At the same time, they bring a little warmth back to the nest, and now all of them were on their feet, ready to colonise more land.

In fact, they show some similarities to bees when it's time to establish a satellite community. Ants, too, send out scouts to locate possible sites, and once they've laid scent trails they compare all the different suggestions. It was unclear to me how exactly this happened, but it hinted at a democratic society.

Which new areas of the property they had set their sights on was obvious. One was the area surrounding the septic tank. When I lifted its wooden hatch, ants carrying eggs and pupae swarmed around the tank. Both the hatch and the frame around it needed a coat of linseed oil, so I stood there for a moment, pondering how to solve this problem. The ants, however, wasted no time. They immediately began to carry the eggs and pupae away in caravans to get them out of the bright light, and in working together they did in fact manage to lift the entire nursery to the top of the wooden frame and push it all inside. Within half an hour, the area was ant-free.

That was an easy solution. Relieved, I replaced the lid. But the next time I lifted it the ants had taken over the entire area again. Pupae and eggs were neatly arranged back in their old spots, until the light streamed in. Then they repeated the same procedure as last time.

Their keen organisation was impressive, almost as if they were each a small part of one big organism. And then it struck me that those two words were related. An organism is a system that can organise itself, all on its own.

My appreciation for the ants' organisation took a blow when I discovered that they had also invaded the writing nook. I didn't discover this right away, since it was in an out-of-the-way corner, but there was a developing ant path, a constant trickle from ceiling to floor. From a distance it looked like Chinese characters, although I suppose any alphabet can be likened to tiny curlicues on their way to creating greater meaning. I have no aspirations to master all languages, but I was quite annoyed at my inability to interpret the ants' message in the very writing nook where I wished to engage with language. It was no help that I had come to form some understanding of the bees' dance, for ants have their own systems of communication.

What were they *doing* in this little outbuilding? It certainly seemed like they were merely following one another. Maybe it provided a sense of comfort for near-sighted creatures who can only see a few centimetres ahead and are better at making out movements than shapes. Ants are, however, capable of navigating on their own. Since their muscles remember how they've moved from one place to the next, they can always find their way home using the position of the sun. And they can mark their path with pheromones, the scent substance produced by their glands.

All organisms have pheromones, and, as we have seen, bees use them to call for help or give encouragement. But in the case of ants, these chemicals seem to have developed into a system something like a language. Since each pheromone has a particular effect, they can be combined to relate even more kinds of meaning, and when they are released at certain intervals it's rather like Morse code. One ant researcher, who deciphered upwards of twenty pheromone-words, even suspected there might be a type of syntax or sentence structure.

What's more, the simple messages of pheromones can have different tenses or levels of intensity based on the time that has passed since they were emitted. A few are quick calls that vanish rapidly, but others are long-lived orientation trails, and these are especially refined. Ants returning from a discovery only leave scent trails if they're carrying something. No scent trail indicates that there's nothing more to gather.

Context and strength can also add meaning to pheromones. A warning signal near the nest provokes aggression, but released further away it becomes an exhortation to flee. A faint signal is a request for additional workers, but a strong signal functions as an attack alarm. If the molecular structure is adjusted, the resulting messages can become secret codes that are only understood within that particular colony.

What's more, the pheromone language can be combined with sound and movement. Some ants can make a creaking sound by rubbing a ridged section of their abdomen; others can shake rhythmically; still others can click their jaws or mandibles. To emphasise something they can gently strike other ants with their antennae, and if the nest is attacked they can easily raise the alarm by banging their heads against something with resonance.

And it's not just that these various signals are broadcast – the way other ants receive them is also quite sophisticated. Pheromones are captured by the antennae, each section of which interprets different scents. One perceives the scent of home and another is for reading scent trails. A third section can tell the age of other ants, while a fourth is tuned in to the aroma of the queen – the one who gives the ant colony its identity. Home, novel paths, the character of fellow ants and an ant's own identity – all of these ring out at once, like a chord.

Like bees, ants use their antennae for both smell and touch. Together they create a nearly three-dimensional picture, like a landscape in relief, made of long or short, tightly packed or loosely spaced shapes. This is crucial, since ants move through a terrain that is like one big odour map. It's full of the smells of different bacteria and fungi, of insects that might be predator or prey. They can also orient themselves using sight. For instance, researchers have noticed how an older ant, accompanying a younger one, stops here and there along the way to let her inexperienced sister find landmarks such as small pine buds or the shadow underneath a bush.

The clouds of pheromones surrounding ants are not merely lightweight speech bubbles, so perhaps there was an important exchange of information going on right next to me. Surely the ants would describe this place much differently to how I was used to – after all, the senses can give rise to an entire way of viewing the world. As my eyes followed the caravan of ants up the wall, it struck me how much of my own language is determined by sight and hearing. It's built upon visual symbols and audible sounds but can only capture feeling, taste and smell with the help of fumbling associations. Perfumes are characterised through images of alluring, elegant or fresh-faced women, and wines are described through laboured comparisons to everything from sharpened pencils to stables.

Ants have it easier. They can make out just as many aromas as we have in our perfumes and wines, and since they don't need to go the roundabout way through words, they can be more exact about it. When I compared my alphabet with the pheromone language of ants, in fact, my own seemed rather fabricated and abstract – which of course it is.

So there I sat beside a complex language of the senses, feeling left out. Perhaps a few discreet vibration signals had even reached the ants' forelegs? Indeed, they have a type of hearing organ near their knees. I thought of Evelyn Glennie, the deaf virtuoso percussionist who plays concerts barefoot to capture the sound waves through her feet. I myself, naturally, couldn't feel a thing.

Perhaps taste was another driving factor in the caravan on the wall. Ants greet each other mouth-to-mouth so that they can simultaneously share the food in their crops and information about what they've found. It seemed wrong to me to call this generous mouth greeting 'regurgitation'. Instead, I associated it with the origin of a kiss. According to one theory, kissing developed out of the mothering habit of feeding pre-chewed food to babies mouth-to-mouth, and only later became an intimate gesture. Whatever the case, the ants' mouth greeting is both a way to share and a messaging system. In some ants, it has contracted into one physical act: an individual ant shares a message about a food discovery by rocking back and forth with its jaws open, as if to share food.

Wasn't it rather touching that their way of greeting was just like feeding a baby? Didn't that say something about how they took care of each other? Sometimes they even try to care for a friend after its death, until they smell the odour of rotting. At that point, the dead ant is rushed away to a waste storage area on the outskirts of the colony. One researcher transferred the corpse odour to living ants and found that they too were quickly carried off, even though they put up lively resistance. The odour meant death, and that was that – for in the world of the ants, scents convey the truth.

Yet they can consciously use their pheromone language to deceive. Just like humans, ants can lie. Cunning ants can, for instance, sneak into other nests and signal, 'Out, attack!' Then when the nest is empty, the traitors are free to come in and steal larvae, which they raise to be slaves.

Lies are a foul but sophisticated way to use a language – they indicate that the liar can anticipate the reactions of others and in doing so, manipulate those others. Even those who lie for egotistical reasons have moved outside their own sphere of thinking. Lies are evidence, then, that ants can understand how other individuals think.

The more I thought about the elementary language of ants, the more clearly I could see its many-faceted nature. It can give guidance or warning; it can provide information about food or show solidarity; it can understand environments; it can delineate roles within a group. Not to mention that it can lie and even encrypt secret intelligence. When African army ants go on ravaging expeditions, the scout troops leave scent trails that instruct the main corps to hold back, advance or surround a victim. And as if that weren't enough, some researchers report that ant language can even be used mathematically. One species of ant, it seems, can combine its language with the value pi to measure surface area.

It was raining outside, and through the patter of raindrops I could hear hammering. A carpenter was nailing up moulding in the little bunkhouse that would become my sister's bedroom. It was nice to hear his talk, now and then, of lath, tongue-and-groove board, and other down-to-earth objects that create a concrete space. Sometimes I would even intersperse my writing periods with time spent weeding out the contents of the workshop, where the former owners had generously left everything behind. The tools invited me to sort through them, with their practical, unambiguous names and uses. Chisels, pliers, files, drills, nails and screws of all sizes could be arranged in a pleasing order, and, that done, I cheerfully sorted out busted electrical cords, dried-up cans of paint and other things that disturbed my neat system. I have long had the desire to straighten out the hubbub of life, and the workshop could be a stand-in.

Words about life are different. They have fuzzy edges, wide-ranging associations and varying layers, so it's hard to build anything truly stable out of them. These days philosophers solve that problem by making language abstract, but in doing so, of course, you're peeling away life itself. Naturally it takes a certain amount of distance to get a complete view of any topic. That goes for me as well, and in order to write about broader contexts I must be alone, away from the distractions of the social layers of language. This is why I've often sought out-of-the-way writing nooks.

The farthest I took this urge was in my romantic youth, when I believed that life's big questions could be dealt with in the course of one summer. I also thought of myself as a woman who loved islands, and I was on the hunt for the most isolated ones I could find. When a travel company launched Robinson Crusoe Weeks on the west coast of Sweden I did not hesitate to contact them. The company would arrange for a tent, supplies and transport by boat to an uninhabited skerry where you would spend a week alone. That was what I thought I longed for: an islet on the horizon of freedom.

On the boat there, I learned that I would be the only one for this Robinson Crusoe week in two senses. It had thus far only been tested out by a former war correspondent who hadn't been able to endure her stay on the skerry due to some very bad thunderstorms.

After the boat left me on the island, I took stock of the necessities I'd been provided with. Besides the tent and a jerry can of water, there was a shapeless dry bag that mostly seemed to contain canned goods. It was a preview of the pleasures awaiting me for the week. Because it was so heavy, I left it all on the beach as I familiarised myself with the island.

It was truly the opposite of civilisation, for it belonged to no one and was about as inviting as if it were home to a recluse. The trees were more like bushes, squat against the elements, and poking up among the roaring breakers were sharp precipices and cairns that looked like suitable homes for snakes. At the water's edge lay fragile, flute-like bird skeletons among boards from boats, broken into wing shapes. So many delicate things seemed to have been smashed there. Something incredible must have happened on this windswept island. In the middle of it was a split rock with a handful of sooty streaks like those left by a lightning bolt. I dragged my gear over to it and put up my tent on the miniature meadow that had formed in the centre of the cleft.

And what was I supposed to do then? It seemed a bit abstract to ponder life, so in want of any other task I started in on a simple canned dinner. Among the food in the dry bag was a camping stove, which I placed on a hillock. It was hard to light, and all I could manage to achieve was a quick, reptilian flame that vanished as soon as it appeared. At the same time, a pungent trickle spread down the rocks and onto my hands. The kerosene chamber was leaking.

Resigned, I went to wash my hands in the sea but found myself just standing there halfway down to the shore. At the water's edge was a slab full of seals. They were packed as close as humans on a beach, but even in their lazy state they were watchfully facing the water, ready to disappear at the slightest hint of danger. To keep from bothering them, I quietly retreated.

It was still warm, although the clouds had already been towering when I arrived on the skerry. I had just finished another round of the island when the rain came tiptoeing across the rocks. As I climbed into the tent, a few drops followed. The tent, too, turned out to be leaky.

It rapidly grew dark outside as the clouds settled over the island like a lid. But then it was suddenly bright, and a thunderclap blended with the shrieking of gulls. The lightning must have struck the water. This turned out to be merely a prelude, for soon the loud wind competed with the constant crashes of thunder.

I had never before been afraid of thunderstorms. In fact, I loved to stand by the window at my partner's house in the countryside to admire the show. But this time was different. When lightning flashed, the zippers of the tent clanged like alarm clocks against the poles that were thrust into the wet meadow. I was surrounded by metal, and I could taste it all on the roof of my mouth. Drops came through the canvas, marking time like the infinite minutes of a water clock. The thunderstorm seemed to be searching its way across the island.

By the time an hour had passed I had cold sores in my mouth and was achy under the eyes. What was life? Tiny electrical impulses that gave rise to heartbeats and muscle contractions in bodies that could also be destroyed by electricity. I was freezing. I had sought freedom only to find myself exposed to the elements.

When the thunder returned that night, I felt an intense longing for the tightly bunched houses of the little coastal community. For all of history, being together and belonging to a group has meant security. I imagined that shimmering schools of fish were moving all around the island, bodies close together like drops in a wave, but my only company was a lone ant that had lost its way and wandered into the tent.

I sensed it at the time, and it has since been proven, that even insects can feel fear. That must have been the feeling that filled the ant, beyond the protection of its community.

It hardly seemed likely that we could comfort one another. I was used to interpreting other beings through voices, if not with words then with songs, purrs, growls, howls or hisses. I could also respond to another creature's feelings with a look, an expression or some body language. None of this was possible with the ant, for she was entirely different. Even aliens, in science fiction, are given human proportions and features. They have two arms, two legs, two eyes and a pair of ears on either side of a nose and mouth. They communicate with language-like sounds and experience us more or less as we do them. Tiny, peculiar earthly creatures, however, can seem far too unlike us to make a real connection.

The ant's body really was bizarre. Its naked chitin skeleton had a cool, metallic gleam, and her eyes were not only tiny but they were made up of facets, so I couldn't make eye contact with her. What little knowledge I had about ants had come to me through books and science. The entomologist Carl Lindroth, for instance, had written a children's book about an ant named Emma, and since it was based on facts about ant life my biology teacher had read parts of it aloud. In it, the courageous Emma encounters antlions, slave-maker ants and parasitoid wasps, and eventually she also gets lost, because one joint of her antennae was broken off when a nursery ant pulled her rather carelessly out of her cocoon at birth. Had something like that happened to the ant in my tent? And what did she feel? A few years later, I would see enlarged X-rays of an ant's brain, where different areas had different colours. It glowed like a church window. I would also see an insect's heart beating in another X-ray film. It didn't look like my own, but it pulsed with just as much life.

The ant sat as if paralysed in a corner of the tent, so incomprehensible to her. We were equally minuscule under the darkened sky, and as such she became, for me, the symbol of a bare and solitary existence. The very fact that we were both caught in our own experiences gave us a type of solidarity. We were alone together. At the same time, I felt deep down in my anxious heart that no man is an island. I come from a city of islands, connected by bridges, and it was those very connections that made the city a whole. Those bridges were life, and they extended even across the boundaries of species.

It was nothing new for an author to combine ants and existential questions. Their tiny size can illustrate the condition of being a vulnerable speck in a massive cosmos, and the meaninglessness of the individual is made plain in their incredible numbers. Only the scent of a constantly birthing queen can keep them alive. And that begs the biggest existential question of all: are we humans just the same? Have we ourselves made up the gods that steer our lives?

Questions like these consumed the Belgian author Maurice Maeterlinck, who received the Nobel Prize in literature in 1911. In my history-of-literature past, I had compared his play *The Blind* with Samuel Beckett's *Waiting for Godot*, which may have been inspired by Maeterlinck. Both plays revolve around a perpetual, futile wait for a leader, and in Maeterlinck's play said leader was particularly essential, as those waiting for him were blind. What they couldn't see was that the leader was already sitting among them. But he was dead.

Although Maeterlinck was best known for his symbolic stage plays, he also wrote eminent books of essays on biology. He dedicated the first one to the bees, as he was a devoted beekeeper. When he was asked to write a film script in the 1920s, he actually tried, to the horror of the producer, to make a bee the hero. Still, in his book, he said some fairly derogatory things about solitary bees. In his opinion, they ought to take the leap from narrow-minded egoism to brotherhood. I got a little stuck on the choice of words, since neither 'narrow-minded egoism' nor 'brotherhood' really describes bees, but I understood that his goal was to extol the virtues of beehive life.

Ants were to an even greater extent his ideal, and in 1930 he wrote a book of essays about their lives as well. As a symbolist he could see our own fate as humans reflected in the simplest anthill. After all, we knew as little about the secret of life as ants did. But the symbolism wasn't heavy-handed, and his book was so full of fascinating facts that it sparked my own interest in ants.

He gave a beautiful description of their lives. It began with tiny eggs, almost invisible, that other ants constantly tended by licking them. Perhaps the ant colony's organisation arose as it did for the very reason that their offspring needed constant care, Maeterlinck mused. Similar things had been said about our own societies, and he thought he could almost see human forms in the larvae that emerged from the eggs. Viewed through a microscope, they resembled grumpy little babies with scornful expressions, or sometimes hooded mummies in sycamore coffins. All the eggs looked identical, except for the one that would become a queen.

After she was helped from her cocoon, veil-like append-ages hung along her sides – these were her wings. It was mind-boggling to think that wings were a memory from the winged ancestors of ants. They would override millions of years of earthbound life just for a day, for a crucial moment. During a few minutes' worth of delirious flight, high above the everyday march, each and every queen could mark the beginning of something new.

It happens each year on a very special afternoon between the hours of five and eight o'clock. The sun shines again after a rain that has softened the ground, and the air is saturated with 70 per cent humidity. It's a mystery how the ants know it, but it never fails. Around dinnertime, the anthills simmer with activity as the young queens are escorted to the surface.

Although they know nothing of the sky, their wings lift them in flight. They're not alone. Each newly born queen in the area takes to the air, and so do the winged ant princes who will fertilise them. It's as if they've all synchron-ised to intermingle their colonies and reduce inbreeding.

Who gives the signal? No one, just a primeval sense for the right time and the right weather. Entire clouds of flying ants rise into the sky beneath hungrily circling birds. Like smoke from an invisible, smouldering fire, they fly until night, when the bats come to take those who are left. Only a small percentage of the thousands of ant queens will survive the day, and the situation is even more dire for the males. After mating, those that escape being eaten by birds fall to the ground, where the workers from their former nest can kill them – they've already made their contribution to ant society on this, their one day of life. With a flurry of excitement, life was given a chance to multiply a thousand-fold. But just as night follows day, death comes right on the heels of life to keep it from collapsing under its own numbers.

I understood why this mating flight had captivated Maeterlinck. It was like an existential landmark, closely related to birth and death. The mating flight of the honey-bee is similarly intense, although in this case there are no swarms. Instead, the queens test the drones by rising ever more boldly towards the sky. They reach points high above the normal flight altitude of bees, until they aren't even visible as tiny dots to normal eyes. This moment is the reason drones have the sharpest vision of all bees. They must not lose track of their queen, for only the one who can follow her to the highest point of her celestial flight can mate with her, although it will cost him his life. During the sky-high act of mating, his innards are pulled from his body; while the queen is filled with life, he falls dead to the ground.

For ants, the mating flight is both a test of strength and a dramatic contrast to their flightless everyday lives. Maeterlinck described it as a country wedding, where the queen's shedding of her wings was like her bridal gown falling away. It was a romanticised image, for there is no party afterwards. The queen must quickly save herself and the lives she will bear by burrowing into the loose soil. In reality, she's digging her own prison. There, in the dark earth, she will lie motionless for all the years she has left to live.

She begins by laying a handful of eggs, which she carefully licks with a nourishing, antibiotic saliva so that no bacteria from the soil will infect them. But her strength is waning, so in order to keep going she must eat some of the eggs she's so carefully tended. There are still millions of sperm in storage inside her body, and from now on it is her life's work to constantly lay fresh, fertilised eggs, as regular as a heartbeat.

Millions of stored sperm … I counted silently. An ant queen can live for upwards of twenty years, and even if only a fraction of those sperm were to fertilise her eggs that could mean hundreds of thousands of new ants. No wonder there are ants everywhere. Ninety-nine per cent of the queens who join the mating flight may die without consequence, because those who survive will multiply and are able to keep the ant society thriving uninterrupted. By caring for and protecting their youngest members as thoroughly as we humans do, they help their young survive the dangerous childhood phase that, for so many species, comes with such a high rate of death.

In Maeterlinck's admiring eyes, the ants' matriarchy was the ideal republic that we had never managed to create for ourselves, and perhaps it was possible because all of them truly were sisters. He considered ants to be some of Earth's most honourable, courageous, generous and devoted creatures, driven by a common altruism. The whole colony would be affected if any one of them were to take more than their share of the common resources, so solidarity and peacefulness ruled among ants. If they encountered other ant colonies, it was only to engage in friendly sporting competitions and games.

It was around this point that I realised Maeterlinck was idealising the ants. Certainly, it was crucial to keep the peace within a colony, but territorial instincts are the bottom line when it comes to the outside world. What Maeterlinck viewed as games and harmless sporting events were described by entomologists as territorial demonstrations of strength. Indeed, the behaviour is so ritualised that it's called a tournament: hundreds of ants doing all they can to impress others. They stretch their legs into stilts and prefer to stand on a piece of gravel to increase their height even further. But it's not just a game. When a larger colony arrives to confront a smaller one, it's time to hurry back into the nest and guard the entrance. For as soon as one side has proven itself to be the stronger, the tournament derails into raids and the weaker colony is enslaved the moment its queen has been killed.

There turned out to be a multitude of similarities between ant societies and our own, and I found that fact a bit discomfiting. The reflection ants aimed back at us from their Lilliputian world could be rather discouraging. Our development of larynxes, which gave us speech, and hands, which allowed us to use tools, have been offered as explanations for the success of our civilisations. Yet ants, with neither larynxes nor hands, still developed organised societies millions of years before we did. They can clearly communicate splendidly through scents, tastes and vibrations, and their jaws can grip just like hands. They can use those jaws to pull loads that weigh twenty times their own body weight, and when other ants pitch in they actually look like fingers on a hand working in tandem. More than any other creatures, ants demonstrate that cooperation can pave the way for advanced societies.

Each ant species demonstrates this in its own way. Weaver ants of the genus *Oecophylla*, for instance, build nests of joined leaves. A single ant can fold a leaf on its own by taking hold of one edge with its jaws and the other with its hind legs. But it takes teamwork to put two leaves together. One ant must grip the first leaf while another holds her rear end and a third holds the second ant's rear end, and so on, until someone can reach the second leaf. Between the leaves are crowds of these ant chains, which sometimes almost braid themselves together. And when the edges are finally lined up, it's time for the next problem: the leaves must be stuck together. The solution is to bring over a larva that's about to spin a cocoon. One ant holds the cocoon-spinner in its jaws and moves it back and forth across the leaf edges, like a shuttle, as it releases its sticky threads. In this way the half-grown pupa becomes a living tool. The work continues until the whole nest looks like a giant, silky, shimmering cocoon of joined leaves.

Ants can also use their bodies as a type of construction material. Fire ants crowd together into a compact mass to create watertight rafts. Tropical army ants can join their bodies into giant tents that both protect the queen and regulate heat and humidity. What's more, ants can use materials from their surroundings; the genus *Aphaenogaster* transforms porous leaves into sponges with which to transport liquid food. In other words, ants have proven themselves capable of using tools.

Thus, even tiny insects can build advanced societies, and they had even done so before us humans. By the time we started tilling the soil about ten thousand years ago, ants had already been cultivating for 50 million years and had been involved in a number of other enterprises as well.

In Texas, so-called harvester ants subsist on a particular variety of grass that they cultivate by weeding out other plants. The tiniest blade of grass is, to an ant, the size of a tree for us, so they must also be lumberjacks, and this in a terrain where each pebble of gravel is essentially a boulder. Even more sophisticated are leaf-cutter ants of the genus *Atta*. Their diet is a fungus that they nourish with so many leaves that they must harvest and process them on a massive scale. Each day, thousands of workers head out to various harvesting sites to pick leaves and cut them into smaller parts. With their multitudes and organisation, the ants can defoliate an entire tree in a day or so, after which huge columns of ants transport the harvest to their mushroom farm. Because the leaf pieces are larger than the ants themselves, it looks as though streams of green dots are flowing all on their own across the land, for the paths, which can be a kilometre long, are constantly cleared by designated road-worker ants. Sometimes smaller ants ride on the leaves like children on a haycart, but these are very serious guards who protect the load from parasites.

It all moves like clockwork. Back in the colony, the leaf pieces are taken to hundreds of underground rooms, which given their size and number resemble factories. The ants have even built a ventilation system, because the fungus farm releases carbon dioxide. This could be dangerous for the ants whose job is to chew leaf pieces into a substrate for the mushrooms. Should any leaves prove to have been treated with pesticides, rendering them harmful to the fungus, the harvest workers are ordered to quickly change growing locations. The ants down in the underground caverns are very observant, and they regularly purge foreign species from their preferred fungus. Besides fertilising it with excrement, they give it growth hormones and a type of antibiotic that protects against microorganisms – both of these are substances produced by their own bodies. At the end of the process, waste is handled by older workers who can be expected to die soon anyway. Everything is as organised as in any industry.

And it's not just cultivation that ants have been up to for millions of years. Long before us, they practised a type of livestock management, although naturally it involves very small animals, namely aphids, which excrete a sweet, energy-rich substance when they eat the sap of plants. It's called by the euphemistic name of 'honeydew', although it's not an ant variety of honey but the profane waste product of a plant louse. The ants diligently milk the honeydew by stroking the aphids with their antennae, and they collect it in such amounts as to suggest a dairy farm. It's also apparent that the ants view it as an animal husbandry endeavour, for when ladybirds arrive to feast upon the aphids they're attacked like predators, and when the aphids grow wings they're torn off just as humans clip the wings of poultry. Black garden ants even store aphid eggs in their nests over the winter, so that in the spring they can place them in a suitable pasture.

The larvae of the blue butterfly genus *Phengaris* also produce honeydew, so some species of ants take these larvae home to their nests where they are nourished with the ants' own eggs in exchange for sweet secretions. Over the winter the larvae pupate in the safety of the nest, and in the spring the new butterflies are chivalrously escorted out by the ants who provided them a home.

Clearly ants are willing to go to great lengths to obtain their honeydew. But they themselves actually eat very little, and they can go months or even a year without food, as long as there is some moisture in the soil. They collect the aphid juice in special pouches on their bodies to feed the ant larvae, which can devour ten kilos of it in a summer. The larvae also need protein, and to provide it the ants drag various insects home. Flies, mosquitoes, butterflies and beetles, worms, spiders or millipedes – they all have larvae that can nourish young ants, and each ant colony goes through a million insects each year. If the victim resists, a dose of formic acid will make them comply. Formic acid is so effective that it's even used by beekeepers and birds, although in their case it's to get rid of mites and other parasites. Starlings will even flat out plop themselves down in an anthill to let the inhabitants squirt their acid on them, and sometimes they even take a few ants in their beak to rub them on their feathers. On top of everything else, an anthill is one of nature's pharmacies.

Indeed, ants truly demonstrate a well-organised society in every way, and the sum of it all rather puts humans to shame. Before we'd even arrived on the scene, ants had had agriculture, animal husbandry, tools and an industrial society for millions of years. We humans were not the creators of the first civilisations on Earth. Ants were.

Of course, advanced societies have their price. They must be defended, for instance, so 15 per cent of the individuals in an ant colony might be soldiers. It's clear which ones they are – they have a sturdier build and sharper jaws than others. But even regular old workers take part in battles when it comes to surrounding an enemy. Although they're often old ladies, in an age sense, they fight like bold Amazons. Ovid wrote that in Greek mythology the gods had transformed ants into a particularly war-prone tribe of people, the Myrmidons. And ants certainly have an astounding variety of military tactics. They make use of infiltration and guerrilla warfare, blockades and sieges, storming and veritable extermination; indeed, they even have suicide bombers that can explode and coat the enemy in a poisonous goo.

This all seemed eerily familiar and set my mind in motion. Ants' battles for territory seldom brought lasting results, so entomologist Carl Lindroth wanted to view them from a broader perspective. Since ants lack serious enemies, perhaps they had to depend on one another to keep their population within reasonable boundaries. From a global perspective, their wars can be viewed as a braking mechanism that prevents their numbers from growing excessively.

I pondered this. After all, from a societal standpoint, ants are the animals that most closely resemble us. We don't have any enemies to keep us in check either, for we have decimated every threat against our world dominion and have become the apex predator. Could it be out of some biological need for balance that we are always on the verge of destroying each other and ourselves with increasingly devastating weapons?

These similarities made biologist E.O. Wilson weave ant and human narratives together in a prizewinning novel, *Anthill*. Like Wilson himself, protagonist Raff becomes fascinated by ants on his lone ramblings as a child, and they later take centre stage in his research. In

Wilson's case, this was all thanks to an accident. As a child he happened to get a fishhook stuck in his eye while fishing alone, and since he was afraid to see a doctor his distance vision was permanently affected. But he turned his lemons into lemonade and became one of the world's leading experts on ants. He was the one to determine the role of pheromones in their communication, among other things.

Wilson himself found communication crucial, so even as a renowned professor at Harvard he didn't settle for confining his research findings to academic circles. He wanted to show everyone the life of ants. In just the same way, he allowed his protagonist Raff to craft his dissertation into a Homeric epic about the rise and fall of several ant kingdoms.

The tale begins with the death of a queen in an ant colony, which faces its downfall as a result. When the queenless ants are challenged to a tournament by a foreign ant colony they are defeated and must rush to entrench themselves in their nest. As they shelter there they are eventually forced to eat their own larvae, and when the conquerors finally invade the nest the entire defeated colony is destroyed, just as the Romans destroyed Carthage. The few ants who manage to flee are only able to survive on their own for hours or days.

But no kingdom lasts forever, and even victors must be conquered. Due to a mutation, another colony nearby has lost their sensitivity for territory-border scents. Even the scent of their own queen is so faint to them that minor queens are established in an ever-widening network. Without respect for ritual border tournaments, the immoderate realm soon swells into neighbouring domains and takes over. No other ants can control this super-colony, so its only weakness is its excessive nature. These ants keep hordes of aphids, which suck the life out of the plants, and the ants themselves scare off pollinating insects by eating up their larvae. There are simply more ants than the habitat can support, and they are on the verge of destroying themselves.

At this point, a higher power – from an ant's perspective – swoops in. These are beings who might bring gifts in the form of leftover bits of picnic food, but who also have the power to eradicate them. In this instance, they vanquish the entire supercolony with a chemical pesticide.

Behind each detail in the book is Wilson's own research on ants, although the parallels with human civilisation are clear. For instance, he allows protagonist Raff to see the similarities between the growing complexity of the ants' society and the kaleidoscope of specialists at Harvard. The interspersed ant epic resembles our own historical chronicles and also provides a frightening vision of the future. In the introduction, Wilson declares that his story has multiple layers. It's about the human and ant worlds alike, but its lessons are also applicable to the biosphere and the Earth, where each species must exist in reasonable proportions.

*   *   *

The ants tirelessly continued their caravan across the wall of the writing nook. There was something singular about the way they stuck together. It had become so crucial to their survival that ants on their own were condemned to death, almost as though in coming together they built an organism.

Could they, in fact, be seen this very way? Maeterlinck was of the opinion that an anthill really should be considered a single being. And it does turn out to be the case that an ant colony, throughout its long life, undergoes changes in character just as an individual does. Younger colonies are sensitive and impulsive like teenagers, while older colonies are more stable. It doesn't matter that each ant lives only for a year or so – each one comes from the same queen, and the society on the whole reflects her age. Ants cooperate not just like loyal citizens and sisters, but like cells in a body. Ten per cent of a colony's residents can be lost each day without much of an effect, for hundreds of thousands remain and new life is constantly being born to preserve the whole.

Wasn't this also the case in my own body? Millions of my cells die constantly, and new ones are always forming. All in all, I am crowded with 37 billion cells, and each one is a tiny living thing of its own even though they cooperate like ants. Like ants, then, they must have a chemical system of communication and a division of labour.

It wasn't hard to see the parallels with an ant colony. Ant soldiers had their counterpart in immune cells, which drive off all foreign invaders, and the ant queen could be compared to the endocrine system, which organises cell division, nutrient absorption and blood circulation. Without making any conscious decisions, it organises the work of the cells and provides constant nourishment for a future.

I took a jaunt to the mirror in the writing nook. What I was looking at was a massive colony of social cells. They created my senses and my brain, the parts of me that could take in my surroundings and let my thoughts wonder about themselves. Even though a thousand brain cells died each day, and with them all their connections, I remained the same. Was it by way of words that I attempted to create a definite *me*?

In my body, there was so much going on of which I was unaware ('I'?). Everything had a different format than what I saw around me. Entire rivers coursed through my tiny blood vessels, which together would reach all the way around the world; in my brain, tiny electrical storms raged as communicating neurons feverishly searched for patterns in the world. In all of this invisible micro-life there was such a paradoxical vastness that it reminded me of outer space. There were as many neurons weaving together impressions, impulses and notions as there were stars in the Milky Way. Among themselves, too, they had tens of thousands of connections that formed a network of everything I had ever heard, felt and seen ('I'?). Each individual brain cell was as limited as its neighbours, but together they linked into a network that could reach far beyond their own purview.

The same went for the cells in each organ. I had only to open my eyes and a fresh, massive concert was set in motion. With every passing second, the 125 million light-sensitive cells of my retina sent new impulses to my brain. There, 10 billion brain cells would construct an image before forwarding the impulses to my 640 muscles.

In every part of me, these were astronomical sums. What could be the meaning behind all these incomprehensible numbers? There had to be some purpose. Consider too that these microscopic fragments were made up of even smaller worlds, for cells consist of atoms. Still, most inconceivable of all was the fact that the better part of all of these components is nothing but emptiness – just as is the case in outer space. If the electrons of my cells were compressed the rest of me would be about the size of an ant.

How could all these countless tiny parts full of inner space create an image of the world? Could the ants give us a clue? While they continued their endless march on the wall, I weighed a USB drive in my hand. It contained the half-finished manuscripts of a couple of books, and a lot of facts. If microchips could hold so much information, shouldn't the same be true of ants' brains?

There are important differences, of course. In contrast to the world of computers, the chemistry of ants, like all life, is carbon-based. Nor is ant cooperation programmed from without; it arises from inside the ants themselves, and their connections reach out to other forms of life.

Still, computer researchers have begun to use social insects as models for self-organised systems. After all, in a data processor, seemingly insignificant ones and zeros come together to transmit complex information quickly.

The explanation lies in a pre-programmed decision node or algorithm. A similar thing was probably taking place in my own brain, and it definitely was in the ant colony. There, each individual makes a simple decision without having a complete picture of a given problem. That decision is often based on the behaviour of neighbours, so, for instance, if many ants are streaming in from a food source, that source is judged to be rich. When a lot of small, local decisions come together, the sum may be much greater than one individual contribution. The limitations of each ant are, in fact, an advantage – if one were to act of its own accord, the greater whole might be jeopardised.

No central decision-maker is needed in this kind of hive intelligence. Ants, like cells, show us how a complex whole can arise from the interaction of smaller parts. Just as a thousand is more than a dozen, a crowd in itself has intrinsic value. All of its individual parts taken together can provide a statistical pattern.

One scholar of this phenomenon was Darwin's cousin, Francis Galton. He enjoyed synthesising information from varied disciplines such as anthropology and statistics, and sometimes he found interesting connections. In 1906 he noted that eight hundred marketgoers together successfully guessed the weight of an ox, even though no individual even came close to the actual number. Some guessed too high and some too low, but between them was an average value that provided the correct answer. With a smaller crowd it wouldn't have worked.

Was it possible, then, that out of quantities and crowds new qualities could arise? That was a little hard to absorb. I belonged to an individualistic culture and had devoted countless years to studying the pioneering figures of the Renaissance. On the other hand, I had also written books about the conditions of creativity, so I knew that no new gains were possible without the millions of tiny ant-steps that were taken by thousands of other people. Oftentimes, in fact, it was those very anonymous contributions that made new discoveries possible. That was probably more or less the way my brain cells worked together to create an image of the world. When enough puzzle pieces had been gathered, a hint of a pattern emerged. It could seem to happen suddenly, and in the realms of art and science credit went to those who spotted it first. But any number of unnamed individuals would have paved the way for such a breakthrough. Together they created the conditions necessary for the emergence of this new thing that had been awaiting discovery.

Nor was this really a new idea. Aristotle saw how complex patterns could be formed in the interplay between simple structures or behaviours. These days the phenomenon is called 'emergence' and it appears to be a basic requirement of life. Atoms together make molecules; protein molecules arranged in a certain way make living cells; cells organised in a certain way make organs; organs are organised to create organisms, which themselves create societies. And so on and so forth in all spheres of life where all the parts fit together and constantly create new patterns. They're built from the bottom up or the inside out, not from the top down.

Outside, the hammering from the other outbuilding had stopped. The tradesmen had left for the day, and I was exhausted from trying to understand what the ants had to tell me. I was getting hungry; or perhaps my cells were experiencing a collective hunger. No matter; it was time to eat. It was past five o'clock, but the sunshine was still warm as I left the writing nook, and the air was fresh after the rain. Perhaps I could sit outside to eat?

Just as I finished making a salad in the kitchen, I saw a winged creature on the windowpane. Oh, an ant queen! I recalled Maeterlinck's lovely description of a nuptial flight and cautiously let her out. But then another ant queen appeared, and once I'd also helped her outside I discovered a third. That was odd. How had they gotten inside? I looked around and realised that a number of ants were crawling on the west wall. They seemed to be coming from a dark border at the junction between wall and ceiling. I approached anxiously and gasped. The dark border was made up of swarming ants. They hadn't accidentally gotten into the house at all – quite the opposite; they were trying to get out. It almost looked like the guts of the wall were welling out.

The realisation hit me like a bolt of lightning. All at once, the ants became troops in gleaming armour who challenged all my territorial instincts by violating the border of the wall. There was no time to think. In an instant, the situation and I transformed. What weapons were at hand? The vacuum. As I raised its orifice to the dark border on the wall it unleashed a commotion among the ants. Alarmed dispatches were sent in every direction to find a way to escape disaster, for this was similar to a war of extermination. But I was merciless. Like a fury I searched for ants in every corner to suck them into the dark belly of the vacuum.

Afterwards, I was shaken. A swarm of ants coming into the room was a different story from the solitary bees and bumblebees on the outside of the cottage. This was an invasion. And how did ants come to be living in the wall? They needed the moisture of soil. Was there moisture in the wall? This was doubly worrying.

I pulled up a chair and sat down heavily as I tried to come to grips with the situation. Even if the ants in the vacuum died, new ant queens would be born next year and go on more mating flights, and the cycle would only continue. Thanks to their constant birth rate, I would never be able to ignore the presence, in my very home, of these creatures that pre-dated humanity. A wall ought to be a border with the world outside, and if the wall contained life of its own that border was very porous indeed.

I had lost my appetite, but I still tried to pick away at

my salad. On the wall where the ants had appeared was a reproduction of a late-19th-century breakfast scene. It was summer, and the light was reflected in the porcelain and glass; in the background was a wall of foliage. The dining table had been moved outside to be truly close to nature. But now I knew you could indeed be close to nature indoors as well. Suddenly the kitchen walls seemed as flimsy as the art poster, and I was sure that just behind it was a much more active scene than in the painting itself.

Then I recalled something Harry Martinson had written in a nature essay: it had once been common practice to place entire anthills in the walls of houses, because the mixture of dry, sandy soil and evergreen needles made cheap insulation and was thought to keep vermin in check. There might be a whole score of anthills in a large farmhouse. What would it have felt like to live in a building with the walls full of them? Surely it was simply thought of as insulation. But Martinson felt something more for the life of anthills. As a poet, he imagined how every single needle had been carried in and put in its place by some tiny ant. He also saw the anthill as a realm of its very own, in a timeless tradition. In the long chronology of ants, it could have been the 1,059th realm of the 16,000th series, where each series covered 2,000 iterations of state. In the world of the ants, time was measured so differently, both because they were tiny and because they were ancient. Faced with insignificant but infinite numbers of realms throughout time, Martinson felt the same awe as when he thought of all the stars in the sky.

I was beginning to feel ashamed. In my excitement, I had blown things out of proportion. Tiny black ants are, in fact, harmless – unlike carpenter ants, they can't drill through wood. Surely their society had existed on the property for the vast amounts of time Martinson mentioned. The fact that they were small made them no less worthy. After all, they were of the most common dimensions found among Earth's creatures.

Besides, size is relative. Maeterlink and Martinson both pointed out that the bodies of ants are made up of atoms with orbiting electrons not unlike the planets around stars. From this perspective, ants and people alike exist somewhere on a spectrum between the incomprehensibly tiny and the incomprehensibly huge. We find ourselves in the very same situation, just like the time I shared my vulnerable state on the deserted island with an ant.

When I thought of her, I became even more gentle of spirit. She must have been separated from her sisters as the thunderstorm approached while they were out gathering material for their shared nest. The others would manage without her, but she would never survive without them. She was just a minuscule piece of the larger context that gave her life meaning. 'Little ant,' I murmured.

That was when I recalled that I too had once been called by that name. I could see the setting clearly, because it had been a big part of my life. There wasn't much light down in the research room of the Nobel Library, where I had spent many years writing. It was below street level, and the medieval arches gave me the feeling of a profound era in which I was a vanishing piece of something larger. Above me, life went on, on several levels, although I couldn't hear it; outside the window I could see legs hurrying past on their way to other destinations.

A number of women usually sat down there, and although we didn't speak to each other there was a sense of camaraderie as we each soldiered on with some topic. I could often be found among the foreign-language stacks and could only slowly make my way through unwieldy arguments and thickets of footnotes. Nearsightedly, I sought a path among words that together formed sentences, and among sentences that together portrayed contexts. When I caught the scent of an interesting trail I knew it was only the start of an equally arduous work when salvaging the find, and persistence was a virtue here.

Late in the day, the Spanish attendant might walk by with boxes of books on a clattering cart. He knew I had occasionally worked with Spanish material, so sometimes we exchanged a few words. Since I was often the last one in the room, he started calling me *Hormiguita*, Little Ant. 'Will you turn out the lights when you go, Hormiguita?' Then he went on with his load of books. In our own ways, we were each doing our share, like ants pulling straws to an anthill – for, like most things in life, our work required contributions from many quarters. The efforts of others weren't always obvious and yet we depended on one another.

The ants had vanished from the cottage kitchen. Surely they were still hard at work inside the wall, just as they most likely had been all along, even though I hadn't noticed them before. The invisible life is certainly the most common one, given that the contributions of millions of anonymous beings are behind almost everything, including societies and books. In passing, someone might be fondly called Little Ant, which, when I thought about it, seemed like high praise.

Chapter Four

A Veranda with a Sea View

ALTHOUGH IT SEEMED RATHER STRANGE to have ants moving around in the wall, almost everything else in the house seemed under control when summer came. The rest of the painting could easily be postponed until after the holidays; in the meantime, I would finally be exchanging tradesmen for family members.

Suddenly another way of life took over. Cheerful voices filled the house. My sister's curious grandchildren expanded the world of the property. On their very first day they wanted to go fishing in the sound, and afterwards we solemnly shared a small fish. It was as if this ceremony inaugurated the children into cottage life.

'I finally feel anchored,' one of my nephews mumbled as we walked across the property.

Anchored ... the word drifted in my mind. We shared lovely memories of rented cottages, often situated by the sea. One was even on an island, and we'd had to row to fetch fresh water. I suspected my sister still dreamed of such coastal spots, even though these days she had a back injury. The fact was, a romantic view of the sea flowed through our veins, thanks to our English heritage. Our paternal grandfather was raised by a marine biologist and our grandmother by a naval officer, and the two of them met on the Atlantic when Grandfather was a ship's doctor. Perhaps that kind of thing left a genetic stamp.

For my own part, I had tried everything from dinghies to schooners but had come to realise I was not a gifted seaman. The thundery days on the deserted island had also dampened my love of the sea, so although I was still fascinated by the water, I preferred to explore it in a more tranquil fashion.

Now I was writing about rivers. In time, they had come to suit me better than sailing voyages, for although they are in constant motion towards the sea, their journey takes them through meadows, trees and cities. Along their winding route they have birthed civilisations, watered crops, transmitted impulses and drawn borders – in short, they have shaped history, so researching them took time. Thus, during my holiday I would be setting out anew on riverboat expeditions.

By the time I returned to land, summer was over and the last of my family members had left the cottage. I planned to spend some more time there, working on what the rivers had shown me, as soon as the tradesmen were done painting. But life had other plans – water-related ones, in fact. The painting was hardly underway before the crew boss called me, his voice tense. He had bad news. When they started to paint the north wall, the one facing the sound, they found that it had suffered severe water damage. It was so badly rotted that it had to be torn down.

Water in a river was one thing, but in a house it could only spell disaster. The ants in the wall were a bagatelle compared with the fungus that causes dry rot. Even the tradesmen were shaken, now that they knew that six months earlier they had been working atop a roof that turned out to rest on rather dubious supports. The rotted wall had to go, and apparently negotiation was out of the question. To be done properly, in fact, the demolition should continue until they'd found timber that was definitely dry. 'Although you can't afford that,' the crew boss added.

To get a good view of the situation, I went out to the cottage as soon as I could. The tradesmen had gone by the time I arrived, and they weren't the only thing missing. The remains of the north wall lay in a sad pile alongside the cottage. Perhaps it had given in without much resistance.

I entered the room with three walls and found that the bunk beds had been dragged out. Where once there had been a wall, there was now an ocean-blue tarp. It was a little hard to take in. A house with three walls was no longer a house. It was more like a bus shelter, or like the open storage shed. In any case, it wasn't a place to live in.

In the kitchen was a bottle of cognac one of the nephews had brought. I poured a glass and took it into the room, along with a chair.

It was all so ironic. Here I had been planning to work on my facts about water, and now one side of the cottage *was* water – or had been, until the rotted wall was torn down. When I pulled back the tarp, I saw the sound out there, glittering innocently as though none of its rainy sea breezes had carried moisture to the cottage. With the tarp folded open, I grimly sat down to take in the view.

As the cognac in my glass disappeared, my despondent mood began to ease after all. The meeting of inside and outside in this room reminded me of a veranda. Paradoxically, it was also rather calming – almost comforting, even – to gaze out at the sound. In preparation for my river journeys I had read about how water made its way into philosophies of life. Rivers are holy in India, and Chinese philosophy compares seeping water to the Tao that nourishes life. Among the old Greek nature philosophers, Thales considered water to be the primeval element of life, and he was right about that. Each drop out there had existed since the dawn of life on Earth. For more than three billion years, those drops had passed through seas, clouds and bedrock, and then through plants and animals in a never-ending cycle, until there was just as much water flowing through plant life as could be found in the earth's rivers.

A sailboat was slowly crossing the sound, rousing memories of the boats I'd spent time on when I was young. Although it was only for a few summers, it felt like I had spent an era in close contact with the water. Most of my memories revolved around the slightly larger sailboats that had at once given me the expanses of the sea and a stable deck to stand on.

What I recalled most clearly was keeping mid-watch on an old, schooner-rigged school ship that provided my initiation into seamanship. Night was turning to day and seven bells had been struck on the ship's bell. At eight bells, when the hourglass turned, it would be time for the port-side team to be relieved, for we kept watch in four-hour shifts. Beneath me slept the starboard team, in the bunks of the mess; above me the sky teemed with stars.

Sea-fire. Foam. The cardinal points lived in a binnacle stand where the compass communed with Earth's magnetic field. Under my oilskins I was wearing a fisherman's sweater, and at my waist hung a fish knife from a tarry-smelling rope I'd spliced and whipped. Others were handling the sails, so I kept watch over the stars and the sea.

The sea's proportions were all its own, its borders as permeable as watercolours. Shiny or matte, light or dark, the shifting colours told of wind and weather. Currents brought warmth from distant coasts, and the sun might rise above the horizon like an island of magma or sink like a golden Atlantis. The tides responded to the phases of the old moon, for it had probably been torn from the Earth in a collision with another celestial body and now circled us in longing revolutions. The water communicated still with its dry lava seas, if from a distance.

The waves also spoke with the winds. Sometimes the waters were in such tumult that the gimballed table in the mess swung like a pendulum as it strove to remain level and full of food and crockery. Many on board then lost their appetite and an almost untouched roast was later hoisted up the mast where the fresh breezes would keep it cool.

On shore, the waves instead brought gifts from distant coasts. During a raid on an island I picked up smooth stones in every colour. They told of how the sea has worked away at cliffs and slopes for millions of years, loosening pieces of rock that have been carried by currents and polished by sand. The same process turned the sharpest shards of glass into soft ovals of light.

The grains of sand were the younger siblings of those stones, and they too had a long relationship with the sea. It incessantly ground mountains to rocks and rocks to sand, so with every second some billion new grains of sand were born. Together their layers told of vanished landscapes, and yet each one was unique as a result of being formed during its journey with the water. When Antonie van Leeuwenhoek placed a grain of sand in his brand-new microscope in the 17th century, he saw what he considered fantastic shapes. In his eyes, he was viewing the ruins of a temple with kneeling figures. Later, when sand could be magnified a hundred times more, its grains were found to more closely resemble planets with chaotic terrain.

Even their sizes could affect the individual fates of these grains. The very finest ones would measure time in an hourglass or dry the ink on old manuscripts. Others might be formed by Tibetan monks into holy mandalas that would be poured into a river leading to the sea. Children used more robust grains of sand to build their little sand-castles on the beach.

Wasn't sand the perfect medium to illustrate the breadth of life? Philosophical Heraclitus likened time to a river and imagined history as nothing more than a child building sandcastles. Each needed water to hold it together. For their part, astronomers compared the Earth to a tiny grain of sand in space. Given those dimensions, the Sun would measure ten centimetres in diameter and be eleven metres away from us. The distance to the near-est star would be three thousand kilometres, and space might contain more stars than there were grains of sand on every beach and in every desert on Earth. And new ones were always being created. From this inconceivable cosmos came all the matter and water on Earth.

Being at sea prompts many a thought about life, since Earth is in fact a ball of water. Not only are two-thirds of it covered in ocean; if you take depth into account the ocean makes up 98 per cent of the sphere we live on. So why does the ocean feel like a different world to those of us residing on the other 2 per cent? With its sea stars, clouds of plankton and flying fish, it's almost like a space all of its own.

Everything there is in constant motion. When birds migrate in the spring they cross the seas, reflecting on its surface like silvery schools of fish. Salmon and eels cross entire oceans to frolic in the brooks of their youth. They find their way with the help of the Earth's magnetic field, as well as pheromones and the specific tastes of streams, and they can sense minuscule changes in temperature and pressure. Equally indefatigable are sea turtles, sailing around the world towards the shores where they were once born and where they prefer to lay their own eggs. They arrive like primeval creatures, guided by memory.

Yet ocean life is not just a reflection of life in the sky. It has different conditions that demand different senses. For instance, light moves slower through water than through air and is rapidly scattered, so many fish in the deep provide their own light. Sound, however, travels both faster and further in water, although it cannot be heard above the surface, which separates the two elements like an invisible wall. You must dip an oar vertically through the surface and place your ear to the shaft to appreciate the sound beneath. This is what fishermen historically did in the South Seas and in West Africa, and in the 15th century Leonardo too discovered the method. But only in the 1940s did researchers begin to listen for the sounds of the sea, and they were overwhelmed by what they heard. They hardly knew how to describe all the different noises. There was creaking, clucking, cracking, croaking and drumming. There was bubbling, howling, jabbering, whining, whistling and plopping. It sounded like sizzling steaks, ear-splitting saws or heavy, rustling chains. Where did these sounds come from? It turned out that some fish clack their jaws, some blow out air and some use special muscles to bang their air bladders. Schools of herring can make such peculiar noises that the Swedish navy once tracked them, certain they were dealing with a submarine.

The recording of fish sounds I'd heard personally were equally astonishing. Some reminded me of the echoes of ringing bells, some sounded like a silver spoon stirring the contents of a small glass and others were like the hum of a spinning top. These sounds were like voices from a world both distant and related. Everyone from the tiniest shrimp up seemed to be sending messages to one another. Even the tone might be saying something, for it was deeper in older and larger fish than in younger, smaller ones. A lovesick cod gave a low growl while a haddock rumbled.

Aristotle suspected that fish can converse, and it does seem to be the case. One researcher, for example, learned to interpret fish sounds that communicated 'annoyance', 'warning' and 'combat alert'. In addition, there are many nuances of body language, such as the holding of the fins in a different position, or a change in body colour or pattern. Some fish even have electrical fields that announce their species, age, sexual maturity and personality to a potential mate.

So it turned out humans had been overlooking the communication methods of 98 per cent of life on Earth. The only thing that separated us from this world was a thin layer of water; beneath the surface was a vast network of sound waves. They run the gamut from solos to duets and choirs. Like birds, fish love to sing for their females at dawn and dusk, and the gobies that young cod depend upon for food can't even mate until the females have heard their song. Unfortunately, these days it is often drowned out in the racket from recreational watercraft, so it probably isn't solely due to overfishing that cod have vanished.

On my old school ship the interest centred around whale song. The bright chirping of belugas could be heard right through the hulls of boats, so they were called the canaries of the sea. The songs of humpback whales, though, are duller. Their chants can go on for hours, with certain sections repeated like a common refrain. Today, it's thought that their memory is assisted by something like rhyme, for even when a humpback song has hundreds of elements each whale remembers it on the way to their breeding grounds. But the song is gradually updated over time by way of new parts with a faster tempo. By the time eight years have passed, the entire song has been remade to keep the repertoire fresh.

I thought the transformation of a whale song was a lovely way to measure time. On the school ship, a bell rang each time the hourglass was turned, and watches relieved one another as the nautical miles added up. A whale song lasted about as long as one bell, and then it started over again.

Like birds, whales seem to use songs for both expression and communication. And why shouldn't those in the sea also wish to create beauty? Down on the ocean floor, a small pufferfish was observed drawing beautiful flower shapes with its fins in the sand. As a final flourish it decorated the big sand flower with seashells, which the fish brought in its mouth, for when the artwork was completed it might attract a female. Surely this same creative impulse is found in whales.

Sperm whales have dryer voices, with little glissades of clicks that can be heard for miles. To a human ear it sounds like a single creak, but the whales themselves must be able to make out the tiniest part, for the sounds can function as an identifier, a call to gather or a warning. They can also be a fine-tuned echolocation that helps the whales find their way even at a depth of a thousand metres.

Sperm whales have the largest brains of any creature on Earth. So what do they use it for? What do they think about? No one knew. The only whales with which people had attempted to communicate were captive dolphins, but it wasn't to learn anything about them. Instead, their trainers tried to teach them to pronounce human words, even though dolphins lack a larynx. But when taught sign language they could understand about sixty signals for our nouns and verbs, and with the aid of these could understand around a thousand sentences. Most of these were along the lines of 'Touch the frisbee with your tail and then jump over it.'

I had observed this at a dolphinarium and found myself rather depressed afterwards. Self-centred humans can be so simple-minded. After all, dolphins' lives in the wild demand much greater intelligence than it takes to show off tricks, and of course their communication system is very different from our own. It's suited to an environment with its own set of demands, so we could never master their language. A dolphin can produce seven hundred clicks per second and, based on the information returned by the echo, form an image of an object a hundred metres away. In doing so, it can not only differentiate between materials like copper and aluminium but also tell whether something is alive, and if so, whether it's friendly or aggressive.

Among themselves, dolphins communicate with whistles. Each seems to have its own signal tone, rather like a name, and one researcher described 186 different whistles, which she sorted into twenty categories for different actions. It truly does appear to be its own language.

At close range, dolphins use gestures or touches instead and these can even be extended to other species, for dolphins do not limit their social circles to their own kind. Aristotle described small boys riding on them, and I had myself, on a Greek odyssey, seen them leap friskily ahead of our ship as if to tow it along. They were playing with us and could tell what course we were on so immediately that they seemed to predict it ahead of time. Greek seamen gladly interpreted it as such. Apollo is said to have taken the form of a dolphin when he went to the mainland to establish an oracle, and that was why it was given the name Delphi.

Could dolphins actually have some archaic memory of kinship with us? We do have a common ancestor, although the family tree later branched off in a complicated way. Fifty million years ago, whales were also related to even-toed ungulates and lived an amphibious life on the coasts. So why did they return to the sea? Was it out of loyalty or due to their foresight?

For that matter, why had I myself been drawn to the ocean, and why was I now gazing out at the sound through the demolished wall? I was hardly alone in my fascination with the sea – it's lured many a writer. Poets have willingly made it a symbol while novelists have filled it with fantasies. In the 19th century, the same whales that were expertly hunted for train oil for the machines of industry were described as sea monsters on old maps.

The classic depiction of whaling is Melville's *Moby-Dick*, which stood on a shelf on the school ship. I had paged through it to refresh my memory of how life onboard was described, for Melville built upon his own experiences on various whaling ships, and he had an ability to identify the similarities between flowing water and life itself. In addition, he had read all he could about the biology of whales and was inspired by a true story.

About thirty years before *Moby-Dick* came out, the whaling ship *Essex* was rammed by an angry sperm whale that had seen its family harpooned. First it struck the ship with its tail, then with its head, and finally with all fifty metric tonnes of its body. Before the ship sank, the crew managed to save a few navigational instruments and provisions on a couple of whaleboats, but in these tiny boats they were soon as vulnerable as the whales they had just been hunting. Perhaps more so, since they found themselves in an unfamiliar element. Waves fifteen metres high fell over the boats, and when the saltwater soaked into their hardtack their thirst became unbearable. The men were surrounded by sea, but only 1 per cent of the Earth's water is the freshwater that mammals need to drink.

They were also fearful of other people. There might be cannibals on the nearest islands, ready to make quarry of the sailors themselves. Eventually they came across a deserted island that brought temporary salvation, but they so voraciously cleared it of anything edible that they soon had to go to sea again. And there they finally began to eat each other.

The interest surrounding *Moby-Dick* had, of course, nothing to do with the details of whaling, but with the book's symbolism. For Captain Ahab, who had lost his leg in a previous encounter with the white whale, the creature was as evil as the Leviathan of the Old Testament. But Melville himself didn't seem to see it that way. In fact, some interpreters viewed Captain Ahab as a symbol of the ruthless profit-chasing that would become our downfall.

Other 19th-century authors too wrote novels with maritime themes, but without the personal experience. The romance of the sea was in the zeitgeist, and the creatures of the ocean were still so unknown that one could fantasise freely about them. In *Toilers of the Sea*, Victor Hugo described an attacking octopus as an otherworldly monster, and when Jules Verne let an octopus assail Captain Nemo's submarine in *Twenty Thousand Leagues Under the Seas*, the creature seemed tailor-made for the science fiction themes of the book.

But oddly enough, octopuses and other cephalopods turned out to be even more stimulating to the imagination in factual reports from biologists. There they truly seemed to represent another world, and I was intrigued, because the out-of-the-ordinary can allow for fresh perspectives. Cephalopods even demonstrated something that touched upon the history of life.

Each of the hundreds of cephalopod species has its own peculiarities. The largest is a giant squid fourteen metres long and the oldest – the tiny nautilus – has navigated its way through 500 million years essentially unchanged, adjusting its buoyancy like an ancient submarine. Other species can blend in with their surroundings by suddenly assuming the shape of a flounder, a snake or a piece of coral.

Still, the most remarkable cephalopod of all is the eight-armed octopus. It has three hearts and the blue blood of a noble, nine brains and eight investigative arms – or, if you prefer, six arms and two so-called legs. Each arm is, in a certain sense, a world of its own, with vision cells, touch receptors, a refined sense of smell and taste, and a sort of short-term memory, so octopuses rather test the boundaries of our concept of an individual.

There is no denying they are multi-faceted. In aquariums, they can use their arms to do puzzles, open jars and pull corks from bottles, when they're not inquisitively inspecting objects dropped down to them. They're observant and quick to learn from other octopuses' problem-solving methods; they have a good memory, so they remember which people have been unpleasant and which have given them food. Since they can tell when they're being watched, they are easily irritated by aquarium visitors, which has prompted some to build small barricades out of stone slabs and others to carry around coconut shells to hide beneath. Still others have figured out that if they squirt water onto the spotlight above their tank it will short out, leading to blessed darkness. One octopus threw rocks at the aquarium glass; another managed to escape with the help of his mental and physical agility. First, he contracted his boneless body and squeezed himself out through a gap in the aquarium lid. Then he crawled down to the floor and a drainpipe that led him back to the sea. Other eight-armed break-out kings soon followed. Some took nightly trips out of their tanks to cross the room and slip into aquariums full of crabs. After a proper feast they dutifully returned to their own exhibits, so the staff had no idea what was going on until the whole caper was captured on a hidden camera.

Thus, octopuses could escape from the limits of their present situation by combining their own memories with plans and complicated actions. I found this more exciting than any horror film, because it was evidence of considerable intelligence.

Still, octopuses don't fit into our theories of intelligence at all, as these are rather descriptions of our own aptitude. In many cases these theories are passably supported by our knowledge of whales, but octopuses sabotage them in the most shameful way. We usually say that the prerequisites for intelligence are an extended period of care during a creature's youth, a prolonged and social life, and the ability to adapt to varied environments. Nothing of this is true of eight-armed octopuses. They are short-lived, starving to death while they tenderly care for their hundreds of thousands of eggs, so they don't live long enough to impart knowledge to their young. They're also asocial and each species keeps to its natural environment. Yet the most embarrassing part is the fact that these brilliant creatures belong to the phylum Mollusca. We are said to be more closely related to sea cucumbers than to octopuses. Apparently, intelligence developed along several paths during the course of evolution and manifests in different ways in different species. Just as echolocation has emerged in both dolphins as well as bats, both octopuses and crows are astute. So evolution has no definite goal or peak, only branches like an octopus's arms, and all of them can point the way to some remarkable realisation about life.

* * *

A sudden breeze caused the blue tarp to flutter in the cottage. It sounded like when the sails of the school ship were about to be taken in rapidly by many arms, for the surface of the sail was as large as the deck, and the size of the ship rivalled that of the bulkiest beasts of the sea. The water's buoyancy doesn't only allow huge dimensions in ships. Waterborne animals can also have a massive size, so the blue whale has become the largest creature that has ever lived.

I too had been fascinated by this colossus, although I was starting to wonder whether an unreasonable amount of attention hadn't been paid to its size. The blue whale builds its 150 metric tonnes upon the tiniest creatures of the sea, and while these creatures may not seem as charismatic as whales, the oceans would be dead without them. In other words, life in miniature could balance out the largest lives and carry as much weight.

The blue whale feeds on plankton, which drift with the currents. Zooplankton can be fish larvae, crabs, mussels, sea stars or barnacles, but the most common kind is krill. A blue whale easily eats four million krill in a day, so enormous amounts of them are needed, and their miles-wide schools are greater than any swarms of birds or insects. *The Guinness Book of World Records* has called them the 'largest aggregation of animal life' because they have even been seen from space. A school of krill can contain as many individuals as I have cells in my body.

Could these schools even be considered a sort of superorganism? Krill are always discussed in the plural, since it's hard to make out any detail in their compact mass. While they drift through the sea together, their tiny swimming legs keep a common rhythm like a single breath. Yet a researcher, when studying them, suddenly saw one tiny creature break out of the school to observe him. It was only a few centimetres long, but it gave the impression of being quite curious with its dark eyes and forward-pointing antennae. Eye to eye, two individuals from different families wondered about one another, and in that moment size made no difference.

My own experience with krill had been as food. The fisherman who sold them said they tasted more or less like shrimp, with a hint of crab. But it's not the flavour that sticks out in my memory. In the evening, when it was time to eat in the darkened room, they took on a bioluminescent glow reminiscent of the deep sea.

Krill are, in fact, safest in deep, dark waters. There, fragments of dead sea creatures filter down like a gentle snowfall. Only at twilight do they venture up to the surface to eat their meals of even tinier plankton, which are sometimes so small that several million of them could fit in a tablespoon. They're called phytoplankton from a Greek word meaning 'plant', but they have no roots or leaves. Instead, they belong to the huge kingdom of algae. And they live in a world of their very own.

There's something remarkable about algae. Even though there are hundreds of thousands of species, most of them remain undiscovered, and they can vary widely. Some are microscopic, while others are sixty metres long; some glow like sea-fire, and some, like bladderwrack, spread eggs and sperm under the full moon in June. Some transfer their red pigmentation to shellfish and corals, while others provide Omega-3 fatty acids to fish, and still others release toxins when they decay. Many also live outside the sea, anywhere from moist stretches of land to springs and lakes.

The biologist I once shared my life with was an expert on freshwater algae, and on occasion I came along as he was looking for them in lakes. The milieu was the peaceful opposite of whale-hunting and the romance of the sea. You could row calmly across the mirrored surface, on the hunt for the green foundation of life along blooming shores. Beneath the tiny fish was a glimpse of the lake bottom, and as glistening drops fell from the oars they formed rings in the water. Time stood still. Anytime the biologist lay down in the bow to keep a lookout through his aquascope I had to take the oars. I could see from his bent back how full of anticipation he was as the dredge came up with its random harvest, almost as if he were on a treasure hunt. Gradually he filled the small bottles on the thwart with samples, and when they were later viewed under the microscope, it turned out that many algae truly did display a jewel-like beauty.

Was this how some of the first children of the water looked? After all, algae are more than just the foundation of the oceans' food chain – they are among the oldest living things on Earth. During the millions of years spent quietly absorbing nourishment from sunlight they were also turning carbon dioxide into carbon and oxygen, forming the air we now need to live.

Considering the fact that life arose about four billion years ago we don't have a very clear picture of how it happened. It could have been that energy from a lightning strike transformed hydrogen, ammonia and methane into organic elements that reacted with one another, but it could also have happened by way of volcanic eruptions on the bottom of the ocean. In any case, most people believe that it happened in the sea, in hot springs under it or in a warm lake. Water was a necessary ingredient.

Ancient water clocks kept time by way of droplets, and I thought that was fitting. In time, such droplets could erode rock and fill entire seas. They also provided the backdrop for the dawn of life, when energy and moisture were captured in microscopic membranes. There was a weak electric force, and thus the cells that became the innermost part of all living things were born.

Although these cells measured only a few thousandths of a millimetre, they comprised all the necessities of life – metabolism, movement and communication. That seemed familiar. Wasn't that exactly how life is in every home? People eat to replace nutrients, move about to perform various tasks and communicate with one another or gaze out the window. That was certainly what we'd been up to in the cottage. And suddenly I saw an incredible, uplifting parallel between our cottage and the cells. Like the cottage, cells were little one-room homes with porous walls, and the very fact that those walls were permeable paved the way for a crucial development. A different, simpler cell – by all appearances a bacterium – made its way through them. But it became a good lodger and contributed to the power supply, until eventually the cell could expand its small room.

It was hardly likely that something similar would happen to the cottage, for in the cellular world this transformation took a goodly amount of time. Still, once penetrated, the cell wall opened itself to a new, expansive chapter of life where plants were able to emerge.

But cells are also united in other ways, and one of them can still be seen among the porous sponges of the ocean floor – those creatures who, after their death, often become household sponges. If living sponges are pressed through a sieve they become mere crumbs, but if they're placed in water they will re-form into a sponge like the one that's just been mashed. Those little crumbs really want to stick together. Even tiny pieces that are torn off can form new sponges. Thus, they must have some system of self-organisation.

Somewhere within this mystery, life unfolded. Cells turned out to be just as social as ants. They could exchange substances right through their membranes and communicate with the help of proteins. And this communication deeply affected me, as it was what kept me alive. Like all living things, I too am made up of cells. They have a language inscribed in water, and it is shared by all beings on Earth. Thanks to this, we all carry something of the origins of life within us.

It was starting to get chilly on my faux veranda, so I went to the kitchen to make a cup of tea. As the steam rose in the saucepan I thought of how smoothly water transforms from sea to cloud, rain to ice. It's in everything that's alive and everything we eat – yes, even in the driest biscuit or piece of crispbread. If anything could be called the basis of life, water was truly it.

I took my cup of tea out to the writing nook and closed the door to keep the warmth in. Four walls surrounded me like a little cell, and why not? Life is written in cells. But how to describe life itself? It was hardly a straightforward story. Storytellers as a rule avoid branching off with many tangents, but that's what life is all about. Cells communicate in all directions and mix up the most varied alphabets. One has been assigned 107 letters, for chemical elements; another is for all the chromosomes of a cell; a third describes the four bases that form the infinite combinations of a DNA spiral. I could never capture such vastness in the linear sequences of our alphabet.

Still, I wanted to understand the story life had written. So let's say it began three or four billion years before my birth. Then came a series of chapters, each of which lasted many millions of years. Since some were more eventful than others, I browsed ahead in my mind, stopping again in the Cambrian Period about 540 million years ago.

By that point, the movements of the continental plates had turned the spot where I was sitting into a warm groundwater pond, decorated with formations of coral. These were the exoskeletons of small, flower-like cnidarians who got their pigmentation and energy from a symbiotic relationship with algae. Once a year, during the full moon, they sent out sperm and eggs so that the new exoskeletons of larvae expanded the coral reef, and in other places such reefs became the largest structures on Earth and home to a quarter of the creatures in the oceans.

Trilobites, those prehistoric creatures that bore a certain resemblance to the woodlouse, left an equally great legacy. While they might not have looked like much to the world at large, they may have been the first to see it, for eyes were a novelty at the time. Those of the trilobites had six-sided parts just like the faceted eyes insects would come to sport later, and since the lenses were crystalline calcite prisms, some have been preserved.

It's not that trilobites were the first to discover water. It takes comparison and distance to get a proper perspective. But the trilobites' diffuse panoramic view of the water was predominant during the 300 million years that their thousands of species ruled the seas. At the same time, they died, like all individuals, and drifted with their calcium-rich eyes to the sea floor to join snails and corals. For millions of years, it was all compressed into the limestone that in one brief second of Earth's history became pyramids, cathedrals, roads, fertiliser and toothpaste. Contained within were some of life's first eyes.

I put down the teacup on the windowsill, where a bowl of fossils united marine history and summer memories. Next to petrified coral and brachiopods were a few younger snails with grooved layers in their shells. These were the decorative calendars of their years of life. One tower shell was like the spiral shapes in the marble of our landing that had fascinated me as a child – it oscilliated with the ancient movements of time. But the fossilised spirals in that marble were orthocerida – cephalopods that lived 400 million years ago.

That was around the time the seas began to fill with fish. They, too, came equipped with an important novelty: they had spines to protect the sensitive nerve fibres between brain and body. From these fish evolved all reptiles, birds and mammals. Greek nature philosopher Anaximander suspected there was a connection when he studied fossils a few thousand years ago. And I too had a fossil before my eyes as I pondered life. Hanging near my desk was an exuberant little fish, preserved in stone. It was one of my ancestors.

It turned out that fish would also ease my growing hunger that afternoon. In my hurry to get out to the rotted wall, I had forgotten to bring proper food, but there were some classic pantry staples in the kitchen – a tin of so-called 'Master' sardines and one of herring. Those would do just fine. Both sardines and herring are among my favourite kinds of fish. I knew that this was a predilection I shared with many others, for both are part of our cultural history.

Salted herring was a Nordic staple for over a thousand years, both on land and at sea. For the Vikings it was a provision, and it later became such an important commodity for the Hanseatic League that the confederation completely fell apart when schools of herring fled for other coasts. Now that I thought about it, such schools had in fact become increasingly scarce in recent decades as well, although this time it was probably due to the fishing industries vacuuming them out of the sea. So perhaps I should go with the tin of Master sardines.

Sardines are part of a southern European cultural history that even encompassed their packaging. During ancient times, salted sardines were stored in amphorae, which were replaced by sturdy wooden barrels after the fall of the Roman Empire. And when those were later exchanged for tins, it was as a result of Napoleon's many wars, as his hundreds of thousands of soldiers needed field provisions that were easy to transport. First, they tried canning the fish in glass jars with a boiling water bath, and this was later replaced with iron boxes that had leaden lids. But they were heavy and hard to open, so it was soon their turn to be replaced with tin cans that had small keys for opening – and they were a huge success. Entire ship crews brought along tins of sardines, even while circumnavigating the world, and as a result sardines got to travel further than any other fish.

Meanwhile, British rule was increasing on the seven seas, and they fed their sailors with catches of herring. It was almost as if the Napoleonic wars lived on through the tiny fish. Still, sardines and herrings aren't that dissimilar. In the 16th century, Conrad Gesner had a rather tough time describing their differences, as sardines belong to the herring family. I myself had learned that smaller herring in the Baltic Sea and larger ones in the North Sea go by different names in Swedish – *strömming* and *sill*, respectively – but today the two types intermingle in those seas. Climate change has also caused sardines to move northwards into the North Sea, so the species seem to be getting all mixed up with one another. This was only confirmed when I took a closer look at the Master sardines tin. Its contents were in fact listed as *skarpsill*, or European sprat, yet another relative of sardines and herring. My choice between herring and sardines had solved itself.

As I arranged the herring-sardines from my tin on a slice of crispbread, I tried to imagine their bodies as they were when alive. Size-wise, they were somewhere between krill and larger fish, and in the water they would have gathered in billowing, silvery ellipses like creatures a kilometre long. They would almost blend into the sea, too, for each fish reflected it off thousands of scales. If they twisted in one direction they would reflect the light from the surface; if they twisted the other way they would throw back the darkness of the deep water. Since the fish at the edges of the school were closer to both food and predators, they would have taken turns being farthest out. They also helped each other find plankton and spot dangers like whales, seals, seabirds or larger fish. Each member of the school was in tune with the others, so if a few of them were captured, their neighbours' heart rates would increase as if in sympathy. Perhaps only one in ten thousand would survive to adulthood, but on the other hand, those who made it could live to be fifteen years old – and there have even been cases of twenty-five-year-old herring. Thus there were most likely a variety of fates to be found in a can of sardines.

While I rinsed the alleged sardine oil from the plate, the water flowed over my hands and something inside me responded. After all, like fish, I was 65 per cent water. Since that water constantly had to be replenished, the well was really the most important spot on the property, but the moisture in my body was actually that of a sea. It was there in the salt of tears, sweat and mucus, and the amniotic fluid where I'd spent the first part of my life was salty as well.

It seems I would have liked to stay there. Foetuses are supposed to turn towards the cervix before birth, but I stubbornly rested on my side. Like an old-fashioned telephone receiver on its cradle, I lay there listening to the sounds outside. They were sharper than in the tiny primitive sea I had grown used to. I made my hugely pregnant mother so heavy that together we crashed through a rotting dock. Clearly the sea had a hold on me.

Once I came out into the air, however, I lost my sense of security around water. I only graduated from swimming class when I needed the certificate as a prerequisite to sail, and by then I associated water with a certain amount of fear. Diving was an obligatory part of the exam. I was seized with dizziness on the bouncy springboard and letting go of it was as challenging as leaving my amniotic fluid had once been. Mammals like me could drown. When my body finally sliced through the surface of the water it felt like an existential trial. It was an encounter with an element that could mean both life and death.

My ancestors among the fish, too, hesitated a long time before venturing into an unfamiliar element. While they were establishing themselves in the sea, other life forms were already crawling onto land. Algae were the first ones here too, and gradually they brought a hopeful, green cast to the Earth. Then ferns and lycophytes increased the oxygen content of the air in the Devonian Period. At the same time, the soil was improved by fungi with tough-as-nails eating habits – they could devour stone. As their acids dissolved rock surfaces, their root threads sucked up the minerals.

Like coral in the sea, the terrestrial fungi allied themselves with algae, which shared their solar energy. This gave rise to a new category of plant – the lichens. These, too, made use of softening acids, which eventually created pockets of earth where mosses could grow. The land was starting to be more habitable. Lobe-finned fishes and lungfish slowly began to crawl out of the sea, in the company of small mites and arachnids.

What followed were millions of years of climate change wherein the seas vacillated between rising and falling. Alongside the new coniferous trees, gigantic dragonflies and metre-long millipedes of the Carboniferous Period arose marshy areas full of decaying vegetation.

Then, during the dry periods and earthquakes of the Permian, came a mass extinction in which 90 per cent of all the ocean species vanished. Trilobites were among those who died off, while some thick-skinned reptile species survived. The ancestors of mammals would arise from one of them. Another species, however, brought about the dinosaurs, which would dominate the Earth for 150 million years – as long as the trilobites had ruled the seas. Meanwhile, my mammal ancestors turned into timid, shrew-sized creatures who only dared to come out at night while the dinosaurs were sleeping.

The turning point came 65 million years ago, when an asteroid struck the Earth. Debris from the strike blocked out the sun for months, and more than half of all species died out, including the dinosaurs. But one befeathered species survived, and so did my shrew-like foremothers, who finally dared to come out of their holes.

In the kitchen window, a spider from the Devonian Period was spinning a web. Outside, a dinosaur descendant was warbling in a tree. I went out among the seasoned old pines, ferns and lichens. Beneath them lay sea sediment and fragments of long-vanished mountains. Through the moss crept microscopic tardigrades, invisible Michelin men with eight legs. They had survived five mass extinctions thanks to their tolerance for dehydration, extreme temperatures, vacuums, high pressure and radiation. Amidst all the death and destruction, there was a tenacious vital force.

* * *

Where does that vital force come from? It is passed along by something much smaller than those tiny water bears – it's inside molecules of DNA. Memory and future are linked in a double helix that can reach two metres in length when unrolled, a chronicle of the history of life. The tiniest parts of it have been there from the start.

Everything within is described on the micro-level. For instance, the manuscript of me had fitted in a fertilised egg cell just one millimetre across, and yet that information would take up twenty-five cubic metres of stacked reference books if it were put into words. Not even a third of it would fit in the writing nook. Each new cell was then provided with its own copy. A few cells would form my heart, others my brain, and still others a spine. Inside them, thousands of chemical reactions were at work, and in some miraculous way every cell fell into its proper role. This was made easier by the fact that each had a slightly different surface, like a puzzle piece, and all genes are already present in the chromosomes in the nucleus of the cell. I am, after all, part of an ancient story handed down through 500 million years, although with every copy small variations are introduced.

During my time as an embryo, evolution was fast-forwarded. As soon as the fertilised egg began to divide, it began to flicker like a spinning kaleidoscope. Soon I looked like a sprout, then a tadpole. I had a tail that disappeared and gills that turned into my middle ear, larynx and part of my jaw. All the while, cells added or took away different parts. My hand looked like a shell that developed five branches, and between them a web formed that soon died back so fingers could take shape. It was the same all over my body. Every cell knew when it was time to develop, divide or die, because they were in tune with one another. At the end, almost 90 per cent of all the cells in my foetal stage died to benefit the whole. Only the cancer cells that dreamed of eternal life ignored the others and kept up their own eternal division. Thus I was doubly grateful for all those that died. Both their legacy and their death had created me.

Since the past felt so close, I was taken aback when my phone made a buzzing sound in the writing nook. It turned out to be the crew boss, who explained that they'd had to stop their work but would return the next day. With that, I was suddenly brought back to the present moment. I had, after all, come out to the cottage to discuss the wall issue.

Still, I was glad to have had such a lonely day, a chance to view the situation from a perspective other than the purely practical. A broadened perspective of time can bring a sense of proportion that really puts human trivialities in their place. The catastrophes that beset the Earth were certainly of a different scale than a rotted wall, and they often paved the way for something new as well, although it took time. Even the rain behind the famous biblical flood could, according to geologists, have gone on for millions of years.

On the whole, the history of life is so vast that it must be shrunk down to a week for us to get an overview. If the burning clump that was Earth were formed on a Sunday night, the first life would have been born on Tuesday. These were cyanobacteria that were to have the planet to themselves until Saturday, when the marine animals appeared. But the second Sunday was a hectic time. In the morning, the first plants crept onto land, and a few hours later they were joined by amphibians and insects. In the afternoon the great reptiles took over the Earth, and half an hour later there were mammals as well, although they had to spend four hours living in the shadow of the dinosaurs. The birds arrived around dinnertime. Just before midnight, apes were climbing in the trees, and thirty seconds before the clock struck midnight on this last day, early hominids began to walk on two legs. The entirety of human history took place within a fraction of a second.

Thus we arrived at the last minute on a common journey through life that did not belong solely to us. It was full of the stories of innumerable species, families and individuals. Without them our own history would have been very different, for others, too, were contributors.

The hunt for the aurochs was a challenge, but after this muscular beast was subjugated eight thousand years ago, its descendants became our dairy cattle and draught animals. They toiled in our first fields, and with the farm full people could devote time to tasks other than hunting for food. The population increased, and alongside it grew the need for an organised society. So much depended on precious animals and plants that the Sumerians developed a written language to list them. Horses allowed for communications and warlike conquering, and trade with sheep's wool birthed a market economy where fortunes could be made on herring and whale oil just as well as purple shells, silkworms and ivory.

The most dramatic change came when horsepower was replaced by fossil fuels. These hid memories from early life on Earth, for coal came from decayed vegetation of the Carboniferous Period, and crude oil was the billions of algae, plankton and animals that had been compressed on ancient ocean floors. It seemed, when they burned, that the millions of years that had passed since their time had been squeezed into a massive explosion that transformed the world in an instant. As agricultural societies became industrialised countries, the schooners of the sea were replaced by container ships and tankers, even as the oil that drove it all had to be pumped up from ever greater depths. And sometimes, hundreds of thousands of tonnes escaped into the sea, where it had all begun.

I had seen it happen at close range, along with some old boating friends. A tanker had released so much oil into the Baltic Sea that the clean-up required extra hands, and we wanted to help. We sat in the military helicopter that would bring us to the outer archipelago, ready to fight like soldiers. Through the windows we spotted peacock colours drifting on the sea below, so different from the shades we were used to seeing from the waves.

The rocks of the skerry where the helicopter landed were covered in a pitch-black sludge. The idea was that we should shovel it into bags, but no matter how we scraped with our shovels it stuck to the rock surface. It seeped into cracks and clung to flowers, and we knew that the nesting seabirds would die from the tiniest dab of oil among their feathers.

This was in the 1980s, and a new term had just been minted for the geological era we were living in. The old name, the Holocene, came from a Greek word for 'wholeness'. The suggestion was that it should be called the Anthropocene instead, a name that came from a Greek word for humans. The name change was not an homage. Until this point, the Earth's disasters had been caused by the planet's own convulsions, or by outer forces such as destructive asteroids. Now, however, we were the ones behind a violent and accelerating revolution, with dire consequences for everything from ecosystems to the climate.

There I stood on the black skerry, seeing the future hidden by a shadow. When the sea turned up in my poetry, it was still a result of awe at the drops of water that created life, and at the fish who were so mysteriously drawn to the Sargasso. But lurking beneath the surface were nuclear submarines, and on the bottom was a stash of depleted nuclear fuel that would be deadly for a hundred thousand years. The perspective of time had died of a cancer-like growth.

A few decades after that day among black boulders, everything we had been warned about was undeniably happening. It felt as if the history of Earth had taken a turn when the decayed organisms were taken from their dark realm of death. Once oil became plastic, 15 tonnes of it went into the seas each minute, where it killed millions of seabirds, thousands of whales, turtles and seals, as fragments meandered up the food chain by way of fish. Once oil became energy, 10 million litres were burned each minute, and carbon dioxide built up in the atmosphere, a growing greenhouse, and everything got hotter. Suddenly, a third of the world's population was said to face water shortages. At the same time, the oceans were rising as glaciers melted and threatened other coasts. And we were told that biological species were vanishing a thousand times faster than before, setting us on a course for Earth's sixth mass extinction. As the Earth's master race, we could see our Promethean fire turning back towards us. The banks of the rivers that had birthed our civilisations were marked by clear signs of where we were heading. We were warned that The Flood was coming.

It was getting dark outside, and the boats in the sound had turned on their lanterns. As I walked across the property, I could also see a point of light moving in the sky. It was probably one of the satellites that orbits the Earth nowadays to surveil it or to transmit sounds and images. Some had carried reluctant passengers, such as the cosmonaut dog Laika in her boiling space capsule. Later, thousands of animals took off in satellites – snails, beetles, butterflies, crickets, wasps, spiders, flies, fish, frogs, turtles, mice, apes and cats. Researchers wanted to find out how all these creatures would fare in space once Earth became uninhabitable.

In the kitchen, I found a flashlight to counter the darkness. Somewhere on the shelves of the storage shed should be the sleeping bag that had accompanied me on sea and shore. It was in a crate at the very back, and I was glad for my cone of light when I saw something moving from the corner of my eye. Perhaps it was one of the animals that only came out at night, when people were sleeping.

The wind had picked up, and in the house with three walls the blue tarp began to flap as if it wanted to sail away. When I got back to the writing nook and wriggled my way into the sleeping bag, I was once more reminded of the old school ship. It felt like I was on my way, and in some ways I suppose I was. From a wider perspective, every living creature is a stream between what has been and what awaits.

But as I turned over in my narrow sleeping bag, my perspective began to shrink after all. It had to, because the next morning the tradesmen would turn up again. Could any improvements be made when the torn-down wall was replaced? After all, life had evolved by way of a porous cell wall.

Perhaps my angle of approach had expanded during the day, for now I could turn the problem around in my mind and view the house in a slightly different way. The old cottage sat at an angle to the kitchen extension. That angle had two walls, and if another wall and a roof were added it would form a room. It would only have three walls, but it would do.

Because wasn't a veranda with a view of the water just what we were missing? That way the cottage could be opened up to a broader perspective than inside its own walls. The sea was home to creatures with other senses, other songs. They were different from ours, but in their own way they were just as remarkable. Beneath those creatures, billions of grains of sand gave testimony of vanished landscapes, for if the history of life were as deep as the Mariana Trench, our own era would merely be the foam on the surface. And yet we had managed to pose such a serious threat to life that we must quickly navigate towards a new era. Or would it always take disasters to make us change our ways?

Through the ticking of my alarm clock I heard the wind whining over in the sound. It spoke of uneasy waves in the ocean kingdom I'd come from and still carried with me.

Chapter Five

The Power of a Wild Ground

REPLACING THE ROTTED WALL TOOK TIME. It had become very obvious that all sides of a house are connected, as a piece of the floor had to be torn out as well. There also had to be wiring for the new wall's outlets under it, and once the electricians were at it, they might at the same time run wiring to the various outbuildings.

Their workplace started out dark and cramped. After they wriggled their way into the crawlspace they had to work from a prone position, and their conversations under there were muffled. When those tall boys finally crawled out from the foundation they had to creep down into the metre-high pumphouse to instal a fuse box for the outbuildings. They were probably looking forward to running the electrical cables across the property instead, but that turned out to be the most troublesome part of all. Given the terrain, it became necessary to hand-dig the barren forest soil, and the electricians couldn't do that. They had to settle for laying the cables across the ground in duct pipes. Who could bury them later? The first person I asked refused the second he saw the property. So it seemed like this wouldn't be a quick job.

But electricity is the sign of modernity and would finally reach the smaller buildings. It would also be nice to have a little evening lantern on the property, for the darkness felt like a remnant of the unknown lands beyond civilisation. For that reason, a number of the neighbouring houses left a light on all night, even though it blotted out the stars. Folks preferred to be in the countryside under urbanely illuminated circumstances.

Was this true of me as well? It wasn't necessarily the wilderness I longed for. I was of two minds when it came to my relationship with the wild. As an author, I sought out exacting, elucidating words, but I knew that spontaneous impulses could bring them to life, much like in Carl Jonas Love Almquist's story *Ormus and Ariman*. Ormus was an orderly god who each day organised everything, while Ariman was an unpredictable god who each night changed what Ormus had planned. The result was unsettling, unusual and strangely beautiful.

Even as a concept, 'the wild' was ambiguous. It was sometimes described as what was free or desolate, sometimes as something untamed or violent. Other languages had other associations, as in the French *sauvage*, a word used for wild species as well as for recluses – perhaps they were connected, given that wild animals, too, typically live independent lives.

Thoreau was probably seeking this very sort of distance when he built his solitary forest home at Walden. It was hardly wilderness, for the nearest urban area was within walking distance, but it was still beyond the sphere of a community. He could concentrate in silence on his own thoughts even as he was close to wild animals. Everything he discovered was captured with a tool he himself had helped shape: a lead pencil that had been developed for his father's pencil factory. Everyone there was surprised that he would leave a thriving company to laze around in the woods, but in fact he dawdled as little as any of the animals he spent time around. Like them, he went each day on a solitary hunt that demanded all of his attention. It was a pretty typical life, for an author.

Since an excavator had half promised to show up, I stayed at the cottage with my papers for a while. The days passed and he didn't appear, so maybe something had come up. I am sympathetic when it comes to the unpredictable, because now and then it interferes with my writing. In fact, it sometimes seems that this unknown variable X is at the core of the equation.

It appeared the unexpected had begun to make itself at home around the property as much as it had in the cottage. The sugar bowl I had put out in the spring to lure ants out of the kitchen mysteriously disappeared one night. The same fate befell a small ceramic bird and a pair of shoes that had been left outside the door. Who had taken them? There was no fence around the place, to be sure, but even undefined borders ought to be respected. Even when I was alone it felt as if someone were moving around outside the house.

One day I spotted a strange man walking across the rocky hill. 'Hello!' I called, hurrying up to him. He explained somewhat embarrassed that he was visiting my neighbour and didn't know where the property line was. He had been following an animal path, which he pointed out to me. 'They were here first,' he added. 'You, or your predecessors, built right over their paths.'

He was right, of course. A territory is older than a property line and of a different nature. It's a meeting of space and time where the terrain is marked by memory. The true owners of this place were the wild animals that sensed and lived by every detail of it.

They were also the wardens. It was impossible to ignore the birds' loud territorial calls, and since they were species-specific I could assume there were a number of avian territories on the property. The nuthatches reacted like motion sensors to anything that moved, so their calls were frequent. Surely there were also a number of four-footed animals. I had seen, for instance, a roe buck chasing a doe across the hill.

And, of course, I knew that the squirrel's territorial instincts were strong. One evening she alerted me to someone sneaking by. I was sitting on the half-finished veranda and writing when she began to chatter angrily from up in a nearby birch, her tail flicking. She was facing north, so I looked that direction just in time to spot another reddish-brown tail vanishing towards the public land. That bushy tail belonged to a fox.

All of a sudden, I knew who had been taking things from the property. Naturally, a fox doesn't distinguish between 'mine' and 'yours', but considers all useful items its own, more or less the way we humans do in nature. To top it off, the property was apparently part of the fox's territory, so we were neighbours.

After this sighting, my thoughts gradually began to revolve around the fox. And it was as though my curiosity was reciprocated. Blueberry-studded territorial markings showed up on stumps and stones around the house, and it seemed to me they were edging ever closer.

Since the man who could bury the electrical cables never showed up, I packed my things to leave. The last evening was warm, and once again I sat under the veranda roof to write. My thoughts were far away when from the corner of my eye I spotted something approaching. I looked up and my pen fell to the floor. The fox was coming across the grass. It was big and grizzled with grey, like a wolf, and when our eyes met it took a step right towards me. I thought I could see a smile playing at its half-open mouth.

In the span of a second, I felt my blood pressure rise. Wild animals don't normally come towards people; they prefer to avoid us. Confused, I put up my hand to ward it off, and the fox turned back as in slow motion.

When I finally crept into my bunk in the small hours, I heard fox-yips from down by the public lands. It sounded raw and wild. The next morning, I found the fox's calling card by the door.

The issue of what is 'wild' seemed more pressing to me after the fox's visit. It wasn't only to be found in nature. As a child I associated it with games about the Wild West, although I had no idea why Indians and whites fought or who the Sioux were. A quarter-century later, my life partner at the time and I were on our way to meet them. We were to write a collection of documentary stories on American Indians.

Much had changed since the time of American colonisation in the 19th century, when the land west of the Mississippi was called the Wild West. Before the territory was incorporated into the United States, the masses of immigrants streaming in saw it as nothing but a wilderness to conquer, and when the land was made United States territory in the 1890s, the Wild West era was over. The end was marked by the United States Army's massacre of Indians at Wounded Knee.

Indians viewed the land, like the air, as belonging to all – it couldn't be owned. The Oglala band of Sioux had long moved freely across the prairie, where massive herds of buffalo gave them food, as well as skins for tipis and clothing. The buffalo, therefore, was considered an essential part of the world. But when borders and railroads were drawn over the prairie in the 19th century, the land the Indians had roamed was divided. Within a few years, six million buffalo were destroyed by soldiers, railroaders and white hunters. The broad swaths where the Plains Indians had lived vanished, and instead the government resettled Indians in areas where the soil was considered too poor to grow crops. When it turned out that ore could be mined from these areas, those treaties too were cast aside.

The story behind the United States' Indian reservations is not unique. White colonists have subjugated indigenous cultures since the time of Columbus, and similar events have played out all over the world, including in Sweden with the Sami. The difference was that it didn't become the stuff of popular culture.

In the 1970s, a new generation of Indians in the United States began to protest – both against historical injustices and the corrupt Bureau of Indian Affairs. The Oglala demonstrated by occupying Wounded Knee, where the Indian massacre had taken place, and later the uprisings continued on their Pine Ridge reservation. This was where my writer partner and I were headed.

When we got off at the final stop of the local bus to walk to the reservation, the driver gave us a warning: 'Watch out. Those Indians don't like outsiders, and there's been gunfire at the border of the reservation.' We put down our backpacks and exchanged quizzical looks. Was this a revival of the Wild West? But the lawlessness mostly seemed to be outside the reservation, so we elected to continue.

We arrived just as the Oglala were about to have a general meeting, and the men who approached us were suspicious. Who were we? Anthropologists? Border control consisted of a long, searching gaze. Then we were accepted. We even won their friendship, and it extended far enough for us to be invited to their religious ceremony one evening.

The locale was in stark contrast to highly decorated churches. It was a plain room with the blinds pulled down, since Indian religions were still opposed at the time. Everyone simply sat on the floor around a tray of sand. This represented the Earth. On it were packets of tobacco tied with bands whose colours marked the cardinal directions – east and west, north and south, sky and earth. Holy buffalo meat was meant to have been placed on this altar as well, but prairie dog meat had to stand in.

The sparing nature of the space turned out not to matter, for all life on Earth was connected in darkness. When the lights went out, a shaman's drum sounded for a few minutes before chanting voices came in. They prayed for 'all of our relations', but this didn't signify any one tribe. The Oglalas' ancestors were named for buffalo, moose and eagles, because they had understood their relationship with the other creatures of Earth long before Darwin did. With equal reverence the prayers counted the animals that were usually eaten. Time seemed to slow as the walls of the room gradually expanded to encompass the vast family, as if the prairie were once again wide and open. Soon my legs were asleep, but I was moving through a world that was whole and limitless. For there in the dark I realised that words like 'wholeness' and 'holiness' have a common root. An awesome thought: everything was connected and had the same value.

Then the scent of salvia spread, and a basin of purifying water was passed around for all to drink from. It was time for the sacred pipe. It was passed to me as well, and I accepted it. As I drew in the smoke, I thought of how the mouthpiece had been touched by all the lips that had formed prayer in the darkness. Was also I now one of those who had prayed for the great family of life?

The Oglala ceremony lingered inside me for a long time as something deeper than an exotic memory. It was born of a spirit I found in many Indian cultures. A story from the Pit River Indians of California, for instance, describes a long migration in which father Bear, mother Antelope, son Fox and daughter Quail meet a series of other animals from their large family. One is the old medicine man, grandfather Coyote, who sees his kinsman in unruly son Fox.

Coyote is often a trickster in Indian legends, someone who moves outside an orderly world. Like a joker, he can unexpectedly turn the rules of life on end, but he's not evil. In Indian mythology, the world was not created by any kinglike, vengeful sort of god, but by animals who embodied the mysterious nature of the wild. There the fox too can be seen as a trickster, and in some tribes the fox has been an object of worship. Thoreau understood this. To him, the fox, like the indigenous people of America, represented a more natural life than that of white society.

To him, it was a good testimonial that the fox had retained the freedom of the untamed.

That autumn I left the property to its fate, but when a cold snap arrived in January I wondered how the cottage and the animals were managing, so I made my way there with a bag of birdseed.

On the property I was met with an embroidery of tracks in the snow. Among light little squirrel hops were both fox tracks and impressions left by the splayed hooves of roe deer. Like droppings and gnawed branches, tracks are signs one can interpret, so now I found myself wondering about a very recent past. I had read that roe deer can produce a secretion from their cloven hooves when they suddenly change course. It's a way to warn relatives about something they've spotted. Were there scent messages hiding among these mixed tracks?

Naturally, the property must be full of tracks year-round, although I hadn't seen them until the snow. The field mice must have made their tunnels beneath the crust, and apparently one led to the cottage, for I found mouse droppings under the sink. They were collected in a corner, next to a rag that perhaps had been a mouse mattress, so the toilet and the bed were neatly separated. If the mice had found something to eat there, besides an old kitchen sponge, I'm sure that too would have had its own spot. At the home of my first long-term partner, the house mice gathered sugar cubes inside a sofa, and I once came across a big-eyed mouse in the pantry, where it had just fortified itself with a bit of gorgonzola ahead of the day's sugar transport. She was adorable, but after that the kitchen got a humané mousetrap.

Since mice must eat constantly, they need to have food in their immediate vicinity. That's why they're drawn to well-stocked houses and barns, although this habit has not endeared them to humans. I recalled a tale of how, in a single day, seventy thousand mice were killed in a grain depot – a good illustration of our view of mice.

Yet our own mammal forebears were mouse-like creatures once upon a time. What's more, we share 80 per cent of our genes with mice, as well as certain traits. Mice, for instance, are extremely social, and when they're not communicating with their ultrasonic voices they skilfully read each other's feelings by way of expressions and scents. When mice saw other mice suffering, in one horrific experiment, they clearly demonstrated sympathy.

We humans, of course, neither perceive the ultrasound nor the subtle expressions, but when mice sounds are brought down to a frequency that humans can hear it's said to sound like birdsong. Male mice, like birds, want to impress females with their songs, and it's said that together a pair can sing relatively complex duets. Apparently this predisposition to song is innate, for it is controlled by the same so-called FOXP2 gene that lies behind birdsong and our own speech. When this language gene is mutated, male mice sing much simpler songs that won't attract a mate.

I once lived close to this tiny world, for as a child my sister and I kept a pair of Japanese waltzing mice. We counted them almost as family, and to make sure they would feel at home an acquaintance built a small mouse house out of cardboard. If you lifted the roof you could see how well organised it was inside, with all the newspaper they had gnawed into hieroglyph-like shapes we couldn't understand. And I'm sure they couldn't understand us either. In their eyes, perhaps, we were frightening masters who might reach down at any moment and pluck them from their hamster wheel. Still, I tried to communicate my tender feelings by stroking their fur with my index finger.

My clearest memory is of the day it was hard to replace the roof once we'd peered into the mouse house. No matter how hard I pushed, something was in the way. A mouse neck had become trapped between the hard cardboard wall and the roof. When I placed the dying mouse in my hand, I saw little eruptions appear in its pelt. They looked like the spinning stars Van Gogh painted near the end, but it was my dripping tears that caused the fur to swirl. Mice and people have always had a complicated relationship.

A Japanese waltzing mouse, of course, has a different life than a field mouse. It might become a child's pet or a researcher's lab animal, while field mice live free. On the other hand, field mice are eaten by predators such as foxes, so their existence too is fraught. When I filled a seed feeder in the birch tree and some spilled on the ground, I felt that the mice were deserving of such a bounty.

For myself I made soup before getting into bed to keep

warm in the chilly cottage. After reading for a while, I turned out the lights and listened to the silence, or perhaps just what I called silence. The world outside was probably full of signals that passed me by.

Around midnight I was woken by a screech. It seemed to come from the public lands, and it sounded dreadfully wild. It was followed by mewling cries. I sat up in my bunk. What was going on, out there in the darkness? Apparently, some drama was underway, one of those where separate lives become enmeshed as a result of life and death. Who had clashed this time? Presumably the fox was involved ... but who else? Once silence was restored, my imagination took this sense of uncertainty and used it to paint ugly pictures in my mind.

When I returned to Stockholm, I began to read more about foxes. If I understood them better, I would perhaps also learn something about the evasive nature of wildness. But, in fact, what initially struck me as I read was how we humans have seen them. Foxes were vilified not only in fairy tales, fables and myths, but also in the Song of Songs in the Bible, which urged humans to capture the fox, that destroyer of vineyards. The fox was always described as cunning, even by Aristotle, who otherwise appreciated intelligence. Why? 'Cunning' suggests a hint of under-handedness, but surely foxes have never tried to hide the fact that they were after food. The thing people seemed to disdain most about them was apparently the untamed spirit that was beyond our control.

Of course foxes need to outfox us. How else can they survive? Countless fox hunts taught them to have multiple exits from their dens and to confuse their pursuers by backtracking or jumping into the water. Perhaps the hunt even honed their ability to find escape routes. In any case, today they live all over the Earth, from barren deserts to high mountaintops.

Their physical shape is an advantage in many situations. They can dig under or climb over various obstacles, and in uneven terrain the whiskers on their paws serve as sensors. On the hunt, their long, narrow bodies can run fast and far, or crawl forward in anticipation of a sudden attack. They often capture rodents with a move known as the 'mousing pounce', wherein the fox listens for movements under the ground for a few seconds, then leaps a metre into the air. By using its tail as a rudder, it can come down more or less on top of the rodent's position. If the fox faces north it seems able to plot an exact course using the Earth's magnetism.

Yet the fox's primary strength seems to be its flexibility. Its diet is labelled as 'opportunistic'. This, too, sounds unreliable. Why not look at it as a creative interplay with changing conditions? Surely foxes dream of plump hens, but it's not as if chickens are their everyday fare. A fox must help itself when the opportunity arises and bury the excess for future days. The fox's main food source is actually small rodents, and when these are scarce there are always worms, insects, cadavers, eggs, ground-dwelling birds, blueberries and blackberries. In an emergency even mushrooms, roots and some grasses will do, for foxes have the teeth of omnivores.

Their adaptability has brought them a number of advantages. Wolves need vast territories to hunt their prey in groups, and when human settlements encroached upon their wild domain they were pushed out. This was a double win for the fox, not only because wolves predate upon them – among houses they found many opportunities. It takes a few square kilometres of poor pine forest to provide sufficient food for a fox, but in a city neighbourhood they can live in the lap of luxury. Rubbish bins are gold mines, since people throw out an awful lot of food. Suburban yards have compost piles, fruits and berries, and there are fewer pesticides there than in the countryside. As an added bonus, hunting is forbidden in populated areas, and city-dwellers aren't as fox-averse as rural folks.

It was during the urbanisation of the 1930s that people noticed foxes starting to move into English cities. By the end of the 20th century, Europe was home to hundreds of thousands of city foxes. Since a human environment had been created everywhere, with just a little nature sprinkled in here and there, many animals learned to survive in the dark corners of human settlements. It became increasingly common to see roe deer, hares, moose, beavers and wild boar on the outskirts of cities, until half of the animal species of the northern hemisphere could be found in populated areas.

Not that they were crowding us out. A thousand years ago, we and our livestock made up two per cent of all the mammals on Earth, but eventually those proportions were reversed. Since our own numbers doubled at regular intervals, we and our domesticated animals now represent more than 90 per cent of the world's mammals. Besides humans, the greater part is made up of our billions of cows and pigs, plus a half billion dogs and a half billion cats. On the other hand, lions, king of the beasts, number under twenty thousand – nearly half of all wild animals have vanished in a short time.

Accordingly, a movement called 'Rewilding' hopes to make parts of Europe wild again with support from the World Wildlife Fund. Fish need to roam our waterways with greater ease, and wild animals can find new territory on former agricultural lands. In fact, for each square kilometre of city there should be a natural area a hundred times larger, where food could be grown and waste treated, so it's in everyone's interest to make more space for nature.

I knew all of this. But how could I personally leave space for wild things? Could a fox be considered a freer variety of dog? It really didn't ask for anything but to be left in peace in its territory. Perhaps it could even become a sort of guide? Foxes must be familiar with the natural world they so flexibly inhabit, and when they absorb every possibility they seem to see the way forward before it even exists.

In April I went back to the cottage again. I was accompanied by that springtime restlessness that is best cured by nature, and there was also the possibility that an excavator would turn up. I intended to do some work in the open storage shed as I waited for him, organising things left behind by my mother and the previous owners.

I'd seen, of course, how houses with three walls are different from those with four. In the case of the storage shed, one implication was that many creatures seemed to feel welcome there. Atop one cabinet was a plundered bird's nest, and the stuffing had been scratched out on the arm of Mum's old sofa. I suspected a cat in both cases. When I moved a framed picture that had been leaning against the wall, I found half a dozen little turds behind it. Was this the cat's idea of tidiness?

To ward off further attacks on the sofa I placed a folding sun lounger on top of it, as an irritated marking of my territory. But when I returned the next day I was met by an incredible sight. On top of the lounger was another marker of territory, an equally irritated response. It was smelly and brown.

My mouth agape, I stared at the brazen little pile. This had gone too far! And what was the intended message? This was the territorial marking of a fox. Was the storage shed to be shared by a cat *and* a fox?

When I thought about it, they had certain things in common. Both are nocturnal, solitary hunters with sensitive whiskers, rough tongues and vertical pupils that allow them to see in poor light. Both can steal forth on tiptoe and arch their backs, and both like to wrap themselves in their tails when they sit or sleep. Both can use their paws to fish and play with the mice and voles they catch, and both are climbers – grey foxes can even retract their claws to keep them sharp, just like cats. Although foxes are members of the canine family, they have adapted to a similar ecological niche as cats.

Yet they live in very different worlds. Cats were the first animals that humans embraced for the sake of pleasure. Though they are mousers, they can't be used as hunters or guards or shepherds as dogs can, and they always retain something of their wild nature. Indoors they might tear out the stuffing of furniture, and outdoors they kill tens of thousands of birds. Even so, we're happy to forgive cats for all of this. After their nightly hunt they can relax on pillows and laps, plied with delicacies.

The opposite goes for foxes, even though they too catch mice. The specially trained dogs of English fox hunts are evidence of just how unsportingly they are hunted. I hated all hunting drives. They were a slap in the face to a superior English tradition: 'you support the underdog'. If the fox was an underdog, I would be rooting for it.

But this fox was certainly testing my patience. When I came out of the storage shed I saw something that had previously escaped me. Between the carpentry shed and the hill stood an old pine tree that had been forced to creep along the ground during tough winters. From its horizontal trunk grew moss and even a small spruce, and beneath this misalliance the pine bashfully crossed its roots. Amongst them I discovered a large hole.

It was a classic fox den. They are typically excavated under trees on slopes, and here the crawlspace of the carpentry shed had provided a secondary exit, free of charge. What's more, all the treasures of the storage shed were right nearby, so this location must have had major advantages for the fox.

I myself could not think of a worse place. Not only could this den destabilise the pine, it was at the very door of the carpentry shed, and no one wants wild animals about their feet when they're going in to enjoy all their well-organised tools. It was true, I had been expecting an excavator, but not this one! And, of course, the fox was *allowed* on the property, but couldn't it stay a little further away? Now I would have to block off the hole once the fox had left for the night. There was no risk of it being trapped inside, given the extra exit.

It was already getting dark when I returned to plan the eviction. From a distance I could see something moving around by the storage shed and wondered if it was the cat or the fox. But when I approached, what I saw instead were two dark little bundles dashing in opposite directions. One vanished into a corner of the shed and the other darted down into the den. They were fox cubs.

Our encounter unleashed two distinct reactions. One of the cubs gazed out curiously from its corner, while the other anxiously remained in its hiding spot. I myself backed towards the cottage, because I had reacted just as wildly as the foxes.

* * *

In his tale *The Little Prince*, Antoine de Saint-Exupéry has a fox explain the advent of a friendship. Neither party must take command, and slowly but surely they may win each other's trust. If the prince sat a little closer each day, the fox would be tamed.

Apparently, friendship between humans and foxes is not unheard of. In the early 2000s, a sixteen-thousand-year-old grave was discovered in the Middle East; the man inside lay close to a fox. Archaeologists were taken aback. Graves that contained dogs alongside humans were four thousand years younger. Could a fox have been sufficiently important that much earlier? Important enough to become a companion even into death? Was it because we humans had been just as wild as the fox at that time with our hunter-gatherer lives?

I wasn't interested in taming the fox at the cottage. Once I found out it had been fed by some neighbours I understood why it wasn't shy of people, but I didn't want to give it food or a name. What interested me was the fox's independence. It wasn't looking for a master; it simply obeyed its own will, and that of life.

A few million years ago, the canine family had branched off into the wolf-like *Canis* and the fox-like *Vulpes*. The domesticated dog is descended from the *Canis* side. Domestication may have begun when some wolves were attracted to leftovers from slaughter near sites inhabited by humans, and those who were unafraid got more to eat and had more young as a result. In turn they began to see humans as a resource, and eventually their offspring could be tamed. Since wolves are pack animals, they have a rigid hierarchy where alpha males are never challenged, so even a human could take the position of a leader with a single wolf. That done, any number of *Canis* varieties ensued. St Bernards and poodles, hunting dogs and watchdogs, bloodhounds and shepherds – each one, down to the tiniest lap dog, has wolf genes.

Foxes, on the other hand, sustain themselves mostly on small rodents that can't be hunted in packs, and the tiny morsels an attentive fox can find won't feed a whole group. Does that mean that foxes couldn't be tamed? This question was posed in the 1950s by the Russian researcher Dmitry Belyayev, who developed it into an experiment. He wanted to find out if the domestication of wolves could be replicated in Siberian silver foxes.

The experiment was conducted at a fur farm where cages holding a few thousand silver foxes stood in echoey metal barns. It was no wonder that these animals were aggressive; to approach them one needed protective gloves a centimetre thick. Belyayev's plan was to have his assistant Lyudmila Trut breed the least aggressive foxes with each other, and then breed the gentlest of their offspring. Once breeding had selected for certain features of tame dogs, the transformation was quick.

A mental change in the silver foxes appeared in what was truly an astonishingly short time. After six generations, the cubs began to wag their tails when Lyudmila appeared. They even licked her hands and lay on their backs so she could scratch them. In the eighth generation, they retained not only their puppy-like trust and playfulness, but also developed some external puppy features. Their noses grew blunter, their ears drooped a little and their tails developed a curve.

There were also hormonal differences. The experiment foxes had higher levels of serotonin in their blood than foxes in an untamed control group, indicating that they had become less aggressive. It was later confirmed in a less-pleasant experiment that these changes existed on the genetic level. Foetuses from calm female foxes were transplanted into aggressive females, and once born the cubs sought out humans, even though their untamed foster mothers punished them for it.

After this, the researchers wanted to find out if the very gentlest of cubs would be able to handle living with Lyudmila. This was a step further than training foxes to be trusting. If it worked, it would be a true domestication, which comes from the Latin *domus*, 'house', and is applied only to animals kept within walls or fences. At first, the cub that was brought into Lyudmila's house didn't want anything to do with domesticated life. Separated from its siblings and closed up in a house, it seemed to lose its will to live and refused to eat. When it finally gave in, it sought shelter in its mistress's bed.

And with that, silver foxes had been domesticated through selective breeding. They were later sold for a thousand dollars apiece, for many people thought it was exciting to keep an exotic pet. The foxes were trained to sit up on command; they were shampooed and fluffed with hairdryers; they lay on their backs to be scratched while they whined and wagged their tails. Yet something still separated them from tame dogs. They were still so independent that they were a bit difficult to handle.

Were the domesticated foxes happier than their fore-mothers? Since they lived in a state of eternal puppyhood, they no longer had to be responsible for their own lives, and the risks that come with freedom were gone. But perhaps something else was lost.

I had once seen a fox led straight across a public square, and it tugged creepingly at its lead, trying to avoid all the curious eyes. It still had that wild shyness. And I had seen a film of tame foxes chasing each other around a living room like hyperactive children, for freedom wasn't the only thing they lacked in houses. There they had neither the challenges nor the varied stimuli that foxes live with in the forest. We seek enrichment through sports, games and other safe sensations, but boredom is unfamiliar to wild animals because their lives are always at stake. And they'll do anything to survive. Meanwhile, more and more money is spent these days on anti-depressants for pets.

My encounter with the fox cubs immediately changed my view of the den by the carpentry shed. What I saw unfold-ing before me now was the culmination of a family saga. Foxes mate in the winter so that the cubs are born into the possibilities of spring. Even bonded pairs begin their courtship with playful invitations. The female challenges the male by lying on her back or nudging him with her rear, only to act teasingly unavailable a moment later. When she's finally ready, they are inseparable for a long time, and their mating is so vociferous that humans have been known to mistake their wailing shrieks for assault. What I had heard on that night in January, then, was wild mating calls, and while I had been picturing a terrifying drama the foxes were probably already curled up together in a sweet pile.

They stick together even after the cubs are born, although they hunt for themselves and split up the parenting duties. It was the female who had to prepare a nest where she could birth the cubs and then stay with them, as baby foxes live dangerously. Since they were dependent on her for milk, warmth and defence, the male had to bring her food, but he was not allowed inside. If he showed up late she might yip for him, so he too had a rough time of it.

In other words, the fox family needed their den, so I would let them stay. They would also remain undisturbed for the time being, since no excavator showed up this time either – but now I was grateful for it. As for me, I had to go back to Stockholm for a while, so it would be some time before I returned to the property.

And then I tried to be as discreet as possible. It's been said that foxes can hear a clock ticking at a distance of thirty metres, so in the evening I stuck to watching for the fox family through the window. In the muted spring light, the bog moss glowed pale among the pines, and a meditative atmosphere settled around me. I didn't see any foxes, but the nature they lived in was imprinted on me. Perhaps getting a glimpse of the wild was to open oneself to something unexpected, rather like in poetry? With that thought, I drifted off to sleep.

I awoke at dawn to a thumping sound. When I looked out of the window, a roe deer was tearing leaves from a redcurrant bush so that its boughs struck the wall. I sneaked out to the steps and suddenly went stiff. Right before me stood a fox, who was unaware of both me and the roe deer. It was looking for birds in the birch.

The sight was totally confounding. Surely birds in flight couldn't be easy prey, and what's more I had thought the fox was a large creature, grizzled with grey. But here it was, small and red. The fact that it turned out to be the male also turned my assumptions on their head, because male foxes are typically larger than females. If any creatures challenged the human desire for neat categories, it was foxes. Sometimes they were catlike, sometimes doglike, and they were consistently unpredictable.

Now it turned out they could show up at unexpected times to feed the many hungry mouths. Sure enough, in the middle of the afternoon I saw the female hunting on the hill. Her great concentration gave me reason to suspect she was a successful hunter – each step blended into the next as if she had already taken in all that lay before her. I'd experienced a similar feeling once along the road, when a roe deer had leapt right between the taut wires of an iron fence. The space its body had to fit through was calculated in a fraction of a second. In the wild, each instant seemed to contain worlds.

Yes, there was a lot the fox cubs must learn, and in their own way they had probably already started. When I sneaked into the carpentry shed to get a screwdriver, I heard them bickering and dragging something around in the crawlspace. A different way of life made itself known beneath my feet as I stood among the well-organised tools. It was a playful test of the world's potential.

It seemed that the open storage shed had also become a playhouse, for a few ropes appeared to have been used for tug-of-war – unless the fox cubs were just trying to see how long they were. And soon they ventured further from the den, until one evening I heard them beneath me in the crawlspace of the cottage. Soon thereafter I saw them playing right outside the window.

At first there were only two. The number of fox cubs in one litter is thought to be adapted to how many voles can be expected in the spring, although it's beyond me how the bodies of foxes can predict such a thing. But maybe the vole forecast wasn't so bad, for soon enough a third cub showed up. Together the siblings practised their mousing pounces on beetles and searched the grass as if a treasure might await them anywhere. Sometimes they nosed at something; sometimes they ate something. Worms, I imagined. Supposedly fox cubs grow fastest in rainy areas where earthworms frequently emerge from the ground.

But they didn't merely use their mouths to eat, nip or grab one another. They also opened their maws at each other as if to say something, and perhaps, in their own way, that was just what they were doing. Along with flattened ears and a curved tail, a half-open fox muzzle is said to be an invitation to play.

The hierarchy among them probably had to do with their character traits. The one who preferred to stay in the nest was potentially a small female, for fox daughters sometimes stay with their mothers and help with new little siblings. Then they can take over the territory themselves. But for now, it was the small father who watched the cubs play from a distance, while the mother obtained food. Was she simply the better hunter? The next time I saw the family, she was taking the cubs on a little hunting lesson. First, she led them on a jaunt over to the compost, with the motherly cub right on her heels. As soon as she stopped, the cub jumped up at her, and when she turned around it rolled onto its back, all cuddly. The most independent cub, however, wanted to explore on its own, so I wondered how this hunting lesson would go.

The mother fox was admirably patient, and her communication with the cubs was probably eloquent. Some of the forty fox vocalisations biologists have identified are used solely by the mother. In the den they speak softly, with burbling huffs, and the cubs are gently lured out with a mewing purr, while a low, coughing bark is a warning that will send them scurrying down into the den.

They've also been observed calling for individuals, who respond as if to a name.

So what does fox language sound like? It's as difficult to describe in words as a human language would be. Sometimes foxes bark like a roe deer, sometimes they hoot like owls, and sometimes they chatter like field-fares, while the yelping of their young sounds like the cheeping of seabirds. It's as if they've gathered up all the sounds of the wild in order to outwit any simple description.

But the fox family didn't remain close neighbours. A young gardener would be doing some planting on the property, and he heroically took on the task of burying the cables as well. With that, the fox family moved to a calmer spot. Their new home was most likely located on the other side of the hill, where I later saw the frolicking silhouettes of the cubs. But that the property still was a hunting ground I understood one evening when the mother fox came hurrying by with a baby squirrel in her jaws. After that I plugged up the den with some stout logs.

Yet the clogged hole continued to attract life. One day I found the chunks of wood torn away and the contents of the hole scattered around the pine. It was soft and white. The den had been lined with the fur a new mother fox pulls from her abdomen to bare her teats and cushion her young. When I picked it up I could feel all the tenderness in it. The soil is damp and raw in a hole in the ground, but here it had been dressed in a downy warmth. Perhaps the natural insulation had been complemented with a bit of stuffing from a cushion in the storage shed.

What had happened since the foxes moved out? There were a number of insect legs in the fluff. They seemed to have come from bees, so apparently the hole had been taken over by large earth bumblebees who probably appreciated all that fluff. But who had pried the logs out of the hole and dug out the lining? That must have been someone who wanted to get at the bumblebee nest. Who loves honey and can tolerate the defensive stings of bumblebees? Yes, a badger.

I should have recognised the signs from the start. It was no fox who had rooted in the earth, leaving behind the shells of devoured snails. Certainly, it's true that both foxes and badgers live off underground creatures, but they have different personalities and habits. Badgers aren't as agile as foxes, so they must compensate with thoroughness. They dig up rodents just as they do roots, and their nocturnal food-finding rounds are done near-sightedly, their snouts swinging back and forth like a metal detector. They happily slurp up worms with a special spaghetti technique, but all sorts of other edibles go down the hatch as well, from slimy snails and toxic toads to angry wasps and bumblebees.

From the distance afforded by books, I was vaguely familiar with the badger lifestyle. As the featured speaker on one of the summer radio programmes I had even played their bubbly mating sounds. Like foxes, badgers communicate with sounds we associate with other animals

– they can cackle, grunt, chatter, growl, yelp, snort, hiss, scream or whimper, depending on the situation. They can also purr like cats to their young, or chirp like songbirds or coo like doves. A frightened or wounded badger can squeak, howl or emit a heartrending wail.

Their extensive repertoire of expressions notwithstanding, badgers have a pretty reclusive nature. Like foxes, they search for food alone, and when badgers meet they generally either ignore each other or grouse. Conflicts are quick to arise between dominant individuals; males will bite each other in the rear, while females go for the face. Yet they stubbornly live together, for there are advantages to sharing a burrow.

I knew where the local badger den was. It lay in a rockfall off towards the road, and from a thicket of wild roses nearby I had once heard a roaring sort of hiss. It was not what you expected to hear among roses, but judging by the sound I was not welcome to investigate.

The den, or sett, was thought to house eight badgers, and was thus medium-sized. Other setts could be several hundred years old and have up to forty underground rooms, connected by tunnels. Shared dormitories are only used during hibernation, when everyone has to keep each other warm. In the warmer parts of the year, they all want their own private space – like in multi-family housing. Compared with the simple dens of foxes, these setts are magnificent.

The well-to-do badger home Kenneth Grahame pictured when he wrote *The Wind in the Willows* was certainly a comfortable one. Kind Mr. Badger could invite his friends Rat and Mole to sumptuous dinners and offer them comfy guest beds, and along one of his corridors was a living room with a cosy fire. This children's-book idyll doesn't quite match up to a badger's life in the wild.

True, a certain level of comfort is necessary in the sett, because badgers spend three-quarters of their lives down there. But in lieu of cosy fires, the cold walls of the burrows are covered in moss and leaves, and in some cases they're even said to have been wallpapered with found sacks. Since the soil is hardly sterile, both bedstraw and bedrooms are switched out at regular intervals, and when winter is over there's a big spring clean. Those at the top of the hierarchy roll up kilos of grass, while others remove any remnants of food from the sett. The floors are then covered with fresh grasses and ferns, and if there's any fragrant wood garlic at hand it can help keep vermin in check. Should one of the badgers have died during the winter, an earthen wall is quickly erected around it to make a grave. The common latrines are located slightly apart from the sett, preferably near the edge of their territory so the musty odours send a message about the members of the group to strangers.

It must be said that badgers prefer to be tidy. Yet they can't get rid of all the underground parasites that find their way into their fur, so an important part of badger life is scratching oneself. Since the others won't help, badgers are forced to perform acrobatic manoeuvres that loosen up their limbs before the night's wanderings, and as such even undesirable little bed mates have a purpose.

There was something about the down-to-earth badger that aroused tender feelings in me. Linnaeus considered them bears, and people later believed them to be related to skunks. But in fact they are mustelids. Though they don't share the litheness of the weasel or the aquatic life of the otter, they are able to climb and swim when needed. Still, their true home is in the earth. It's their element just as the air is the birds' and the water the fishes'. Roots dangle like broken electrical wires in their nests, for their greatest protection is darkness. As humans spread out, the wilderness shrinks; we have become ever more difficult to escape, and more and more animals flee into the night. The coyotes of California, the brown bears of Alaska, the leopards of Gabon and the lions of Tanzania have all begun to be nocturnal, and this is also true of the elephants of Kenya, despite their poor night vision. The night is becoming their final outpost.

Within the night is a frightening enigma. When darkness falls, we turn on lights and close the door against what takes over outside – the eyes of predators glowing in the shadows, sounds we don't understand. Primeval-looking bats use echolocation to see hidden worlds, and woodlice that breathe through gills creep forth to transform something wilted into soil. They are all at home in the very darkness we flee from. But because they're also part of life, we might encounter them sooner or later.

Of course, there's also a poetic element to a summer

night. When the ground cools, the Roman snails seek green leaves or a partner with whom to perform their gentle mating. Scents linger among calming breezes, brushing by the antennae of moths. Dusk might be followed by the purring flight of nightjars or the cascading notes of nightingales. It's a delight to spend nights like these outside.

The gardener had just buried the electrical cables, which would transfer their energy underground. I myself had remained in the cottage with my papers, so I could lean into early-summer poetry. Something inside me must have responded to the wilderness, for my senses were sharpened beyond the social buzz. As I wrote, I thought I could hear the sound of wings or a low rustle. I looked up from my papers. What was going on out there?

A gentle light drew me out, and on the steps I was met by an apricot moon. Then I discovered something blinking at my feet. I bent down to pick it up and found a female glowworm in my hand. Her emerald-green taillight had just illuminated to guide a male.

Between the moon and the light of the glowworm, it was an enchanted evening. Even the rocks seemed to have come alive, for one of them was moving slowly behind the blueberry thicket. But wait – that was no rock. Something that looked like an African wizard mask appeared.

I was more curious than afraid. The banding of light and dark reminded me of the slice of each day when nocturnal animals relieve diurnal ones. This was where we found ourselves, the badger and me.

Some have interpreted the striping on the badger's face as camouflage to blend in with nature. Others have seen it as a way to appear threatening, as a badger will hide its face to show submission. Biologists have also categorised a dozen physical positions badgers assume to address situations such as danger, defence, invitations or aggression. But the badger before me neither ruffled up its fur nor crouched down, and the eyes that met my own were calm.

What a remarkable creature there was before me. Although badgers have been persistently hunted, there's a lot we don't know about them, and their two-toned pelts reflect a paradoxical nature. They're independent, but they live in groups. They're shy, but they will bravely defend themselves and their families. They're nocturnal, but they have poor vision, so most of those we see are bloody lumps at the side of the road. Yet I found myself sharing the evening with a pair of living eyes.

To make eye contact with a wild animal is to challenge it. Our eyes met with unambiguous curiosity. Perhaps I myself was an enigmatic sight in the moonlight, or perhaps my posture indicated that I wasn't frightened yet, only surprised. Some such encounters only become possible once you relinquish control, and in this case we had both done so. Only when I moved did the badger back cautiously in among the rock and blueberry bushes. That heavy, rolling motion was part of the presentation.

This encounter would have become unavoidable, sooner or later: I had come to realise that the paths through the blueberries had been made by badgers. As creatures of habit they stick to their well-trodden paths, and what's more one of them led to the writing nook and down into its crawlspace. We had unknowingly been sharing the path; it was ours during the day and theirs during the night.

But late one evening, when I was absorbed in my papers, that boundary of time was crossed once again. Deep in thought, I was leaving the writing nook when two badger cubs unexpectedly popped up in front of me. Our view of one another had been blocked by the corner of the building and the blueberry bushes, so none of us were prepared for the encounter.

There is no playbook when it comes to an unexpected situation. The surprise might unleash small fireworks of impulses, as they did the first time the fox cubs found themselves faced with me, and the same thing happened now. One cub awkwardly tried to back up on the narrow path, doing the best it could with its pudgy badger body. The other, though, approached with nearsighted interest. Apparently, it had not yet learned about perils of life nor the brawly nature of badgers. For my part, I cautiously retreated to the writing nook and told myself I was doing so out of consideration. We were on their side of the day, and there was probably a badger mama nearby who might misunderstand this whole thing. It would have been enough to stomp my feet if I felt like moving on, but instead I closed the door and left the night time in peace.

Two moths were dancing stubbornly against the windowpane. My desk lamp was reflected in their eyes, so after a moment I turned it off and lay down in the bunk. The badgers were probably rooting around for underground life out there, but almost everyone else was asleep since our side of the planet had turned away from the light. Squirrels snoozed in the trees; fish slumbered out in the sound. As in the Indian ceremony I had once been permitted to share, many lives were connected in the dark.

Dreams can create inner scenes in the most varied of realms, twitching the legs of both fruit flies and dreaming dogs. One researcher who watched a sleeping octopus change colour believed it was catching a crab in its dreams. At night, after all, different worlds can collide even among the convolutions of our brains. When sensory input is dulled, the traces of the day may be collated undisturbed, allowing playful images to unfurl. Moving pictures can make their own connections in three-dimensional layers, far beyond the rational line of A to B. Oftentimes the night shows me solutions to the problems of the day.

Now I wanted to dig down into the layers beyond language, for I was tired, and tired of thinking in words that kept me awake. Where were my dreams? How do you lure them in? I thought I could hear steps in the darkness, but that was my own pulse carrying on through the night.

At last I fell asleep for a while, until a thudding sound woke me up. It arose neither from my heart nor from any leaf-eating roe deer. It was the badger cubs topping off their wild night out with a wrestling match against the wall. They stopped when I peered out the window, and one began to nose at the doormat instead.

I smiled to myself. A writing nook with dark badger tunnels beneath it was almost too obvious a psycho-analytic symbol. But it had never been my own psyche I wanted to explore; rather, I was interested in what I had in common with others. Yes, even with badgers. Although I lived with words, my brain was guided by the same mute processes as those of the wild ones outside. The expression of the subconscious has at times been called intuition, or instinct, but it's far more than automatic reflexes. It's something as ancient as life itself, and it can still give rise to something new.

Perhaps at its very deepest point a creative life shares something with animals of the night. There is a fox-like openness to multiple possibilities and a sensitive ear for every fleeting nuance, although even that can take a badger-like nature to detect. The wild is many-faceted. It is shy and bold, solitary and playful, and it is responsive to those humans who search it out. After all, we too are children of the Earth.

Chapter Six

The Guardian Tree

I HAD UNDERSTOOD ONE THING AT LAST: the quiet nature of the property was deceiving. Life and communication were always around me, although most of it passed me by. It was telling that I had always encountered the animals around the cottage when in an attentive state of solitude. Upon the return of my lively family, the creatures carefully kept their distance or became background noise.

The plants were a different story. They were always among us, for it was in their foliage we found our holiday energy. Trees supported swings and flowers accompanied us to the kitchen table, and we were happy for them to multiply. Before the electrical cables were buried, the gardener had planted various things that were now beginning to show promise. Abutting the steep drop to the north were protective bushes of cinquefoil and lilac, and some honeysuckles were climbing on the south side. While a cultivated raspberry bush was wilting, sweet wild raspberries were spreading up from the public land. It was more or less the same for the grass that was meant to replace the soulless gravel lot. Those blades that reluctantly poked up were joined by mosses, hawkweed, campion and small patches of orpine that were self-seeding and suited this ground.

Like the plants, the animals demonstrated a will of their own. The squirrel feeder we set up was ignored, perhaps as a show of independence. The same went for a little bee hotel that had been installed on the south-facing wall. The wild bees flew sovereignly by, choosing instead to expand the door-frame nest into a window casing. They also had their own ideas about greenery and would, without a doubt, have disdained a monoculture lawn. They were, however, enamoured by the orpine, which was spread eagerly by the bumblebees.

In fact, boring lawns were not to my taste either. In the 18th century they were status symbols in front of palaces, but now they surround every house. In the United States grass lawns take up three times the surface area of all the cornfields in the country. Their upkeep devours billions of dollars, millions of kilos of insecticides, and the better part of their owners' fresh water.

It was actually liberating that the property preferred to take care of itself. Ever since Adam and Eve were driven out of the Garden of Eden, their descendants have dreamt of a paradise of their own in which to work by the sweat of their brows. I myself have never been particularly fond of weeding, although I know there can be a lot of love behind keeping a garden. In fact, it was the garden my sister created at her house that enabled her to put down some roots in a new country.

Fruit trees would hardly have thrived in the forest ground of this property, but there were pines, junipers, oaks and birches. The lushest birches were on either side of the cottage, one in front of the door and the other at the north-eastern corner. The corner birch grew so close to the cottage that one branch tried to embrace it, while the roots heaved up a couple of stone slabs. Come autumn, the gardener would probably have to prune it back a bit.

The relationship between tree and house has always been a close one. The timber for walls, floor and ceiling holds memories of trees, and wood can lend a room a snug warmth. In the old days, it was a Swedish tradition to plant a 'guardian tree' next to a house so that the roots would suck moisture from the foundation even as the tree's soul protected the house. Perhaps our birch saw itself as a guardian tree?

In any case, it was around the birch we gathered, for it was next to the veranda. If the weather was warm it was pleasant to sit under the shade of the veranda roof and there was also room for a large table. When my sister arrived with her youngest children and grandchildren, three generations managed to squeeze around it, and since there were only three walls to the veranda, nature could join in on the action.

Now and then it even gently honed in on what we were doing. Previously, the kids' older cousins had been playing with snails. Now these children found a grasshopper they named Ferdinand, and it temporarily resided in a moss-lined bowl. At this point, I recalled that my sister had once told stories about an ant with the same name. And one of her sons shared her ability to make up stories. My favourite was about a troll who was so slow that it seemed a little dumb, but as soon as it placed its hands on the moss it was given the answer to any question, because the moss had been around since the beginning of life on Earth. I thought it was a very clever troll, to understand that plants have things to tell us.

A lot came back to me on that veranda. Games came out in the evenings, just as they had when my sister and I were little, and during one game of Memory I had a sense of déjà vu. Everything started over again, in new versions and new generations. It felt rather like the way trees form year rings as they put out new shoots.

Each spring, trees revisit their magic trick of turning sun and water into foliage, and yet I am just as astounded every time. When even the venerable pines are full of spring feelings, they spread pollen everywhere. I'd heard that a hundred million grains of pollen would powder only a single square metre, and I didn't doubt it. Even those grains that landed on the roof and windowsills radiated a belief in the future.

Yet it seemed to me the most eager spring fever was that of the birches. I understood why they were associated with the Norse goddess of fertility, Freyja, for birch sap is said to be rich in energy, and the leaves are tied to the old fertility rites of Midsummer, which marked the end of the spring carnival. It began with leaves like little sequins, and when the sloe blossoms were frothy as a boat's wake, the present and imperfect lived a double life. It was now, and it was a moment ago. Soon the foliage would darken in the light.

Plants know all about the relativity and enormity of time. They can pack it into tiny seeds and make it last forever. They have been incessantly wilting and reborn during a hundred million years, and they still represent 99 per cent of the biomass on Earth. These proportions made me ponder. They meant that we humans, along with all other creatures, make up only a mere fraction of life. There could be no doubt that our planet belongs first and foremost to the plant kingdom.

Plants are also part of most things around us, whether they've been transformed into walls or heat, clothing fibres or tools, medicine or paint – and above all, they are behind everything we eat and drink, since meat animals too live on plants or plant-eaters. Moreover, every breath we take is full of the oxygen they produce. If there's anything we should try to understand, it's plants.

I had begun by seeking out their personalities and shapes with the help of names. Even among the grasses I discovered a world bursting with diversity. It was at once fragile and so robust that a red fescue plant can live for a thousand years. To ants they must be like forests: the tufted grass their pines, the quaking-grass their aspens and the colonial bentgrass their birches.

Also living on the ground were those seeds that were the result of the springtime rendezvous between pollen and pistils. I was moved by the parental care given to them by plants – before tiny seeds are released into the world, they receive both provisions and instructions for handling a number of situations. Biologist Thor Hanson had filled an entire book with the adventures that awaited seeds.

The nourishment with which they are supplied is, of course, desirable to many animals, but this is part of the plan as well. Some seeds, like nuts, have hard shells, and others have an off-putting taste or contain a toxic element. All seeds must also have the ability to travel. Some are embedded in sweet berries or fruits, the better to be transported in animal bellies and deposited in a pile of fertiliser. Others are equipped with tiny hooks so they can hitch a ride on an animal's fur or a bird's feathers. But most seeds have their own wings, propellers or parachutes. In this way, trees have at least a memory of flight.

While most seeds don't go far from their parent plant, the wind carries some a long way. Seeds have been found high above the treeline in the Himalayas, and among those that are carried on water, the fluffy bolls of the cotton plant have made it all the way across the Atlantic.

Then comes the opposite of flight: stillness, the ability to wait. For what is a seed but a vessel for the future? When the British Museum was bombed in the Second

World War, rain fell through the ceiling and suddenly some three-hundred-year-old seeds began to sprout from their herbarium page. Their parents had lived in a different time and a different part of the world, but the seeds themselves were open to new possibilities. Centuries can be mere instants to seeds. They have even sprouted after spending thousands of years in an Egyptian grave.

During their Sleeping Beauty phase, they aren't entirely unaware of what's going on around them. They wait for signals that indicate good conditions. From within their tiny world they even seem to tell seasons, for they can awaken during fires as if responding to a spring-like warmth.

Seeds appeared to be shrouded in so many mysteries. How could they sense light and darkness, heat and moisture? How did they know so much about the Earth and about time? How did they figure out what their embryonic leaves and roots should do once the time was right? Indeed, how could the experience from millions of years be packed into such a tiny seed?

Strictly speaking, it was thanks to the seed's ability to wait that we were sitting in a cottage, eating bread. The seeds enabled our forefathers to become farmers who planned crops and had permanent homes. As gatherers they had lived hand to mouth, and even as settlers they seemed to have eaten soundly of everything they found. The food in an early Syrian settlement was shown to consist of 250 different kinds of plants, for they lived in an expansive green pantry. It's still there, for those who seek it out, and when I found a cookbook for foragers I took the opportunity to try it out with my sister during one summer holiday. Curious, I made creamed orpine, lamb's-quarters patties and pancakes with yellow bedstraw in the batter. So handy to have food right outside the door! My nephews ate it all with a mixture of fear and wonder, sneaking suspicious glances at me. Could you really eat this stuff? Their scepticism grew when I informed them that macaroni and wheat buns are also made of a type of grass, and when I said that candy actually comes from sugarcane plants that grow three metres high, they thought it was a tall tale.

Certain species of grass certainly had taken over the ancient farmers' diets. And as a result the plants were transformed, becoming as rooted as the settlers. As such, people began to claim property and adorn it with new plants. Meanwhile, the number of plants people ate began to shrink, until at last it mostly consisted of cereals, chick-peas and lentils.

Oddly enough, the same pattern appeared in different civilisations. Barley, wheat and rye began to be cultivated in the fertile crescent of the Middle East ten thousand years ago. Around the same time, rice was sown in China and maize in the Americas, and in Africa durra and millet were all the rage. In the end, 70 per cent of all cultivated land was devoted to grasses like rice, barley, corn and wheat. Genetically modified wheat brought even better harvests, although it also took more pesticides and artificial fertilisers to grow.

But seeds weren't only to be found in cereals. Others became spices and medicines. The strong flavours and toxins that were meant as protection against insects became extremely sought after among humans, because proportion is everything. In India, spices were used both in food and in Ayurvedic medicine, and they were similarly cherished in Mesopotamia and China. Then their triumphal march continued into Europe. In ancient Greece, taxes could be paid in black pepper, and in the Roman Empire, nutmeg was used as currency. And as a new era dawned, ships began to search the world around for even more spices. Columbus crossed the Atlantic to find new paths to them, and Vasco da Gama sought a better passage to the spice islands of the Indian Ocean. The hunt was on not only for flavourful seeds but for the saffron of crocuses, the vanilla pods of orchids and the bark of cinnamon trees. The Dutch East India Company made such profits off the spice trade that it would be comparable to today's oil industry. Even the newly discovered plants of South America would turn out to be worth more than the gold of the Indian territory, although people were initially sceptical of potatoes, strawberries, tomatoes and maize.

Rare flowers soon became equally sought after, which is how speculation on Turkish tulips led to a stock market crash in Holland. In certain places, too, the cultivation of exotic plants became linked to the increasingly lucrative slave trade. Africans were called wild savages, and what was wild could be treated with impunity. Thus, both enslaved people and plants were transported to plantations on the other side of the ocean. Though sugarcane is a grass with razor-sharp leaves, it eventually became the world's most popular crop. Equally successful was the cotton plant, which is actually related to hibiscus. In Mexico and South America it had been cultivated by the Aztecs and the Inca, and in Europe the cotton plants Alexander the Great brought back from India spread along the Mediterranean. But it was a novelty in the southern states of North America, and with the forced labour of enslaved people it became the world's first mass-produced material.

The aftermath of this great plant migration was still apparent where we sat on the veranda. Turkish lilacs flourished nearby, and there was cotton cellulose not only in our clothing but in ice cream, margarine, chewing gum and makeup. Like billions of others, we enjoyed our daily coffee, which is now one of the leading goods on the global market. As is also the case with tea, chocolate and tobacco, the wealth accumulates not where the plants are grown but somewhere else, and the coffee requires so much water that it takes 140 litres to make a single cup.

Plants can have peculiar links to history and culture. Caffeine causes the neurons of our brains to react faster, so it's said that coffee paved the way for the Enlightenment. Voltaire and Diderot, at least, drank around forty cups a day at cafés where the discussions were as sober as they were lively. Today, every Swedish employer is more than happy to provide coffee in the workplace.

In fact, caffeine is the plant's protection against insects, and a spider that lands in a cup of coffee will weave strange webs. But nothing is an absolute, and even coffee bushes need pollinators, so small doses of caffeine can attract certain bees without scaring them off. I imagined that they needed perking up after their industrious work.

The exotic plants I'd spent the most time with were houseplants. Plants brought indoors are like quiet pets, and sometimes they show us how they lived in their homelands. Geraniums, for instance, are used to African drought, while Christmas cactuses blossom more or less as they do in Brazil. The prayer plant comes from the same area, and closes its beautiful leaves at the time when Brazilian night begins to fall.

But most of our trees have domestic roots, and they often seem to lend a sense of belonging to a place. Just as Carl Linnaeus's name stemmed from a linden tree on his family farm, whole forests of other trees have cropped up among Swedish surnames. We have Björk and Ek, Lind and Hägg, Rönnblom and Lönnblad, Hasselgren and Almkvist, Alström and Asplund, Furuland and Grankvist. That is to say, Birch and Oak, Linden and Bird-cherry, Rowan-flower and Maple-leaf, Hazel-branch and Elm-twig, Alder-stream and Aspen-grove, Pine-land and Spruce-twig. Perhaps having a tree name was a way of anchoring oneself. Could there be a similar purpose behind the concept of the guardian tree?

My own extended family, of course, had deeper roots. Although the family was spread out, it felt nice to share memories and traits within it. The family tree I tried to draw for the youngest members was rather limited, since its branches soon became tangled. It extended over half a dozen countries, so the roots were sprawling, and a hundred years back there were thousands of connections. A tree depicting our biological evolution would be even trickier to draw, since in that case we were merely a species in a genus in a family in an order in a class in a phylum in a kingdom in a domain of the big family of Earth.

Still, heredity is part of identity. At the same time it's open to change, for even if a cowslip will never be a rose, there is interplay between genetic traits and the environment. How these variations actually surfaced remained a great mystery until the middle of the 19th century, when a man began to plant peas to figure it out once and for all.

Cultivation wasn't actually what Gregor Mendel intended to devote his time to, even though his family had a small farm. What interested him were questions about life. Thanks to scholarships and income from private lessons he was able to study philosophy, physics and mathematics, but it took a fortune or a patron to be able to conduct research. Mendel had neither. His professor therefore advised him to apply to an Augustine monastery that supported scientific study.

Monasteries had been cultivating medicinal plants since the Middle Ages, and when Mendel was given a teaching position he was also allowed to take a few university courses. Thanks to these and a microscope, he was able to tackle the question he was struggling with.

Farmers had long been cross-breeding different species without understanding how their genes transferred, and Linnaeus's classification system left no room for crosses or hybrids. So Mendel wanted to get to the bottom of it by way of experiments. He tried using bees at first, but this was a failure, since bees mate in the air and have fairly peculiar genetics. His attempts with different-coloured mice were more promising, but the abbot didn't approve. Mice have never been welcome creatures, and their habit of multiplying was far too bestial for a monastery.

Plants were a different story. A side benefit of peas was that they produced food for the kitchen, so Mendel was given a greenhouse of his own in the monastery gardens. There, undisturbed, he could matchmake his friends in their flowerbeds, and in eight years he managed to pollinate more than ten thousand pea plants with a tiny paintbrush. He crossed green peas with yellow, wrinkly ones with smooth, white-blossomed ones with violet, and tall ones with short, and thanks to his mathematical training he was able to keep tables of what he found. But he was surprised at the results. When he paired tall plants with short ones, the result wasn't medium-height plants but tall ones; and a cross between white- and violet-blossomed peas gave only white blossoms. Traits that didn't appear in the first generation might show up later, however, so they seemed to be there all along, if hidden. Just as there were two parents, the dominant and recessive genes must come in pairs.

Mendel published his tables in a local scientific journal and sent one of the hundred or so copies to Darwin. He'd read a German translation of Darwin's theory of evolution and had taken detailed notes. But the copy he sent to Darwin would always remain unopened, and interest remained low among Mendel's other contemporaries as well. Compared with Darwin's theories of our animal forefathers, Mendel's pea tables were really quite boring. Thus his findings, like seeds, had to await a new era and only emerged into the light after his death.

When, at last, attention was paid to Mendel's theories, genetics became a branch of science all of its own, although mysteries remained. How could the great, messy diversity of life come out of such well-organised genes? This problem would be investigated by a different person with a different plant.

Barbara McClintock was born just after the 20th century dawned and Mendel's laws became widely known. She, too, had to choose between having children of her own or researching about the heredity of offsprings, for women were dismissed from university when they married. Then again, McClintock hardly had time for a family. She spent 16 hours a day on her research, splitting that time between the laboratory and the fields where she grew maize. To her, maize plants were more than just a research material. She was their Cupid, hurrying from male flowers to female flowers to fertilise them before the wind could beat her to it. Moving through the field, she nearly disappeared, because despite her nickname 'Big Mac' she was shorter than all the plants. When she looked at the maize cells under her microscope she also felt like disappearing – but in a different way. The work was so absorbing that she almost became one with what she saw, and her attentive gaze resulted in the discovery of more details than others had discerned.

In this micro-world, she was entertaining big questions about inheritance and change. In Mendel's mathematical tables, genes looked like neatly strung pearl necklaces, but her own results showed something wilder, more random. Some genes jumped around in incomprehensible ways.

Regular patterns and irregular ones obey different sets of rules, and accordingly they must, to some extent, be understood from different perspectives. McClintock was so attentive to this fact that at last she managed to follow the jumping genes that can transform an inherited trait. But her contributions would receive just as little attention as Mendel's tables once had. People still didn't believe that plants could tell us anything about ourselves. Only under the electron microscopes of the 1970s would other researchers see what she had argued some thirty years earlier. Fragments of chromosomes could move around. This explained the jumble of variations in all species and eventually earned her a Nobel Prize.

By then, her research had also contributed to cultural history. For ten thousand years, maize had accompanied the indigenous people of present-day Latin America, and the time-markers she found in its chromosomes could be compared to the way indigenous societies succeeded one another. Thus, the plant cells under her microscope not only mapped the winding roads of heredity – they also showed the cultural history of a continent.

After all, it seemed rather logical that plants provided the explanation for the inheritance of life, as depicted in the form of family trees. What's more, behind them all stands a world tree with roots all over the Earth. It's found in the mythology of Polynesians, the Yakut of Siberia and the Oglala Sioux, as well as in the Upanishad texts of India. The Babylonians even had two trees: the Tree of Knowledge and the Tree of Life, which were later adopted by Jews and Christians.

Everyone agreed that there was knowledge to be gained from the branches of a tree. Buddha found nirvana under an Indian bodhi tree, and Zeus was said to answer life's questions by way of the wind's rustling in the sacred oaks at Dodona. The messages of more northerly oaks were interpreted by learned Druids, while Vikings communed with their gods in a more brutal fashion in sacrificial groves.

The world tree is described in detail in the Old Norse *Edda*, where it was called Yggdrasil. It had three roots, each anchored in its own spring. At the first sat the Norns, who spun, plied or cut the thread of life. They were comparable to the gods of creation, preservation and destruction in the Upanishads. The second root of Yggdrasil was nourished by the well of Mímir, where the waters held a knowledge of everything that had happened and was yet to come. The third root, however, was surrounded by a cold hell, where it was incessantly gnawed by a snake.

It can be difficult to imagine a world tree with cosmic dimensions, even if you live in it. The people who lived in the middle of Yggdrasil – that is, in Midgard – didn't even understand that they resided in a tree. It's true that their forefathers Ask and Embla were said to have been created from pieces of wood, and the runic alphabet that could grasp the events of time had been carved into bark. But the only ones with an overview of Yggdrasil were Odin's ravens Huginn and Muninn – thought and memory – and they sat in the crown of the tree. With its many branches and runners, it was rather reminiscent of all the synapses of a brain.

Yggdrasil is said to have been an ash tree, just like the classic Swedish guardian tree, but taxonomy wasn't so important until Linnaeus got involved. For instance, stags were said to munch on its needles, so one of my author colleagues guessed that Yggdrasil was actually a yew. I myself felt that most indications pointed to its being a birch, for that was the first tree that came to Scandinavia after the Ice Age. Several traits in the description of Yggdrasil were also familiar to me. Its crown and the ground, for instance, were joined by the squirrel Ratatosk, courier between heaven and Earth, just like the squirrel in our own guardian tree. Supposedly stags grazed in the vicinity of the tree, and given the unreliable nature of species identification in ancient times, these could just as easily have been roe deer like those on the property. And certainly Yggdrasil must have had leaves, for bees were said to be tempted by a dripping dew of honey, and that, of course, comes from leaf-feeding aphids. Since one thing can give rise to the next in trees, they contain a multitude of life, from the birds in their branches to the world of their roots, where earth bumblebees, ants, field mice and foxes might live.

From a scientific standpoint, the image of Yggdrasil as our original home isn't unreasonable. About 3.2 million years ago, for instance, one of our foremothers fell out of another tree. She could walk somewhat upright if she needed to, but down on the ground she was helpless among swift hyenas and sabre-toothed tigers. The trees provided refuge, and both her strong arms and nimble fingers were well adapted for climbing. The only thing she lacked was the easy physics of a squirrel, for she herself weighed almost forty kilos. Despite her relative heaviness, she had made it twelve metres up in the swaying crown at least once. One might wonder what she was doing up there. She came to be called Lucy by researchers, a reference to the Beatles' song 'Lucy in the Sky with Diamonds'. That was about LSD, so maybe there was some narcotic fruit up in the tree? Was Lucy literally high? In any case, she lost her grip on a branch. Because of her weight, she picked up speed as she tumbled down – up to 60 kilometres per hour, which was way too fast. Her injuries indicated that she had time to put out her arms to break her fall, but it was of no use. When the tree no longer held her, the Earth itself became her death.

Why have trees been given so much significance in mythology? Is it because of old memories? Children, after all, are often tree-climbers. The graceful branches of the birches on the property would hardly support such use, but the way they bent towards the ground formed a leafy sort of hut. And for me, this truly awakened memories of a life in the treetops.

From the small back garden, an elm managed to make its way up to my tiny Stockholm balcony. Year after year I had monitored its approach, and when the leaves finally reached the balcony it felt like having a tree fort halfway to heaven. The elm was fabulous company. In the spring it brought forth tiny fruits in rounded wings that looked like silver coins; they made a nutty addition to salads. Fittingly enough, some Swedes call them manna. Then the elm leafed out, and every leaf filled with branching veins like the ones in my own hands. And something peculiar happened among them. So that each one could reach for the sun, they arranged themselves very democratically. The crown took on the shape of a ladder, where the lowest leaves grew a little larger than the next ones up – and they also got extra pigment with which to absorb sunlight. The idea was that no branch should be superior to any other.

Not that the hundreds of thousands of leaves didn't have their differences. Besides their staggered placement, they were also formed by a genetic mosaic. Yet they all emerged from the same trunk, and in a sisterly fashion shared the water the tree gave them. On warm days they transpired hundreds of litres of moisture, which benefited others. At night, each leaf fell into a relaxed slumber that made them hang just a bit lower. In the autumn, a few leaves clung to their twigs a little longer than others, and many had time to perform a little dance before they all gathered on the ground. Together they would weigh as much as a fully packed suitcase, and by then they really had travelled for months, with the tree and the Earth, around the sun.

Unfortunately, my close relationship with the elm came to an end one day. An inspector thought its roots might penetrate the building's foundation, so it was decided that the tree should come down. I remembered the time back in the 1970s when an entire grassroots movement arose because some of Stockholm's elms were to make way for a subway entrance. Protesting friends of the elms quickly ensconced themselves in hammocks and tents among the trees, and after a number of dust-ups managed to save them. I was less successful, so my elm was cut down. It turned out to be an unnecessary measure, since the roots had stayed out of the foundation. But among them something more was discovered. The elm's story wasn't over yet. The stump put out new shoots, and I was given a cross section of it, where the tree's history was inscribed.

I had always seen the elm from above. That was an unusual vantage point in itself, and now I could see it from the inside out as well. Near the middle of the disc was a hole left over from an old rot attack the tree had managed to ward off. Around it arced annual rings that told of the tree's growth. On the side that had faced the house the rings were smaller than on the other side, where there had been more space and light. Some rings were also thinner and had probably formed during tougher years. When I counted them all, I found that the elm had just turned forty. That's around the age when elms begin to blossom. They can live for up to five hundred years, if by then their timber has not already been transformed into a beautifully veined table or the bottom of a sailboat.

What can the insides of a tree tell us? A tree's life history is taken into account when it's selected to become a musical instrument. Luthiers building violins prefer to make the top out of spruce that has grown slowly through the changing seasons, preferably with hard winters and sharp mountain winds that encourage the development of strong fibres. The back and ribs, though, should be made of Balkan maple that has grown in a different environment. To create interplay between two types of wood, a sound post transmits vibrations between them; to make sure that no secondary tones interrupt the resonance the proportions must be accurate down to the millimetre. As wood is a living material, the instrument must also be played regularly.

Is something of the tree's spirit expressed by violins, guitars and woodwinds? What did the living trees themselves have to say? Their anatomy is entirely different from our own, so they must communicate in other ways. I had discussed the topic with my sister, who was in the habit of talking to a cherry tree in her garden. For my part, I was rather hoping for scientific explanations, and they did in fact begin to emerge.

Trees are not as lonesome as they appear, for they have entered into a partnership with the largest organisms on Earth – fungi, which can have root systems a kilometre wide. Even in the early history of foliage, they supplied plants with minerals they sucked up from the bedrock, and when they wound their threadlike roots, or hyphae, around the tree's own roots, both parties benefited. The tree shared the solar energy it received, and in return the fungi gave both nutrition and access to their mycelia. As a result the tree could extend the chemical connection in its interior parts to other trees. They gained a hidden network. Eventually most plants would cooperate with the help of fungi, although those that are purposefully bred seem to have a slightly harder time communicating.

In any case, it's clear that trees do care about one another. They recognise their siblings and can adapt themselves in some ways according to the needs of others. If one tree sounds the alarm about an insect attack, its neighbours can quickly mobilise their own defences. Oaks add acrid tannins to their leaves, and those of goat willows get bitter salicylate. Other compounds can lure enemies of the attackers, which then fight off the attack as well. If the damage is natural, however, the tree merely produces healing hormones and doesn't bother anybody else.

Naturally, the fact that trees communicate raises questions. Certainly, contact occurs on a molecular level, but behind any mode of communication there must be some sort of consciousness in both parties. No one has ever quite agreed on what consciousness is or where it is situated, yet neurologists and cognition experts had determined that any creature with a nervous system can have subjective experiences. But that was in the case of animals; what about plants?

This question has popped up at regular intervals ever since ancient times and has been answered in a variety of ways. Democritus viewed trees as upside-down humans with their brains in the ground. Pythagoras suspected that transmigration of the soul included plants and refused to eat beans as a result. Aristotle got hung up on plants' inability to move around and assigned them an inferior sort of soul.

And on it went. The so-called panpsychists saw a consciousness in every living thing, while rationalists like Descartes saw every living thing that wasn't human as a soulless machine; he was influenced by the great invention of the 17th century, the mechanical pendulum that had by then begun to mark time. In the 1700s, the mechanical view of nature was soundly rejected by Rousseau, who, unlike Descartes, spent time in nature and therefore knew it better. Along similar lines, Linnaeus informed the world that plants multiplied and also slept, based on the fact that they changed position during the night. Surely this indicated a certain level of consciousness when they were alert. In the 19th century, Darwin chimed in with arguments against Descartes' mechanical view of nature. For him, the differences between consciousness in humans and other animals were a matter of degree, and he didn't find it out of the question that plants, too, could have some sort of intelligence.

Eventually, technical experiments would enter the discussion. One of them occurred rather by accident, and under fairly strange circumstances. In 1966, CIA interrogation specialist Cleve Backster was supposed to teach police to use a lie detector or polygraph. Among other factors, it registered increased moisture on the skin, and one morning Backster had an idea as he was watering the potted plants in his office. Could the polygraph measure the speed at which the water travelled from roots to leaves? He attached the electrodes to the leaves, but nothing happened. Then his experience as an interrogator kicked in. Could a more aggressive method perhaps provoke a reaction? What if he were to singe the leaves a little bit? The polygraph immediately registered a reading. The alarmed Backster didn't know what to think. Could plants be sensitive to threats? And if so, what kind of communication system did they have?

His continued experiments attracted both attention and opposition. Since, like us, plants have a system of veins, it was not unreasonable to suspect they could have an internal system of communication. The idea they could talk to each other, however, was considered absurd. But half a century later, research from agricultural universities confirmed that both wheat and corn plants can transmit small messages to one another through their roots and through the air. Could plants have sensitivity down to a cellular or molecular level? A new area of research arose and was given the name 'plant neurobiology'.

The centre of this new discipline was the International Laboratory of Plant Neurobiology in Florence, Italy, where its founder Stefano Mancuso wanted to test a hypothesis. Did plants have a type of swarm intelligence like that of ants?

It wasn't an entirely new concept. Maurice Maeterlinck had expressed similar thoughts – alongside his books about bees and ants, he wrote a volume called *The Intelligence of Flowers*, and it wouldn't have been much of a reach for him to compare plants to ants. To be sure, Maeterlinck was not a scientist but a lay writer, and intelligence is difficult to define even in a human context. But a hundred years after Maeterlinck's book was published, Professor Mancuso explored the idea from a more scientific standpoint. To him, intelligence meant the ability to solve the problems presented by life, and technically it could be defined as being flexible in response to outer stimuli. In principle, it could be studied on an electromagnetic or molecular level.

Mancuso was convinced that different types of intelligence had developed during the course of evolution. One is linked to large, gifted brains like those of humans and was reminiscent of supercomputers. Another type of intelligence, however, is spread out like millions of inter-connected computers. Although each individual has a limited capacity, in cooperation they can produce consid-erable complexity. This is how bees, ants and plants can simultaneously be individuals and parts of a greater whole.

I thought of my encounters with individual bees and ants, and their advanced societies. Then I compared them to plants. You could tell by looking at the tiny details that made each flower unique that plants, too, were individ-uals. On the other hand, the word 'individual' actually means 'indivisible', and you can of course mow grass or take cuttings of geraniums without killing them. Trees can even become stronger when they're pruned.

But there is an explanation. Plants are built differently to us. For us, decapitation means instant death, but decap-itated insects can live for a little while and plants don't have heads at all. When you are constantly being eaten by others and lack the ability to escape, it would be fatal to have all your vital parts in the same place, so the senses of plants are spread out. We don't see them as eyes, noses or mouths, but plants have cells in their leaves that take in light, and their groping roots can feel their way to water and nutrients in the ground. They can even form more roots to absorb what they need. If they sense damaging elements such as lead or cadmium, the roots can retreat.

The above-ground parts of plants move just as much as those beneath the surface. If you snap a picture of flowers every twenty seconds and edit the images into a timelapse film you can see how fiercely they work their way out of the ground. The plant follows the path of the sun in furious little spirals even as it shoots up, and it relaxes only at night. Clinging plants go one step further, reaching for and grabbing on to supports, so if you put a rake next to a honeysuckle plant it will find its way there. A carnivorous plant, one that obtains nutrients from insects, can even feel an insect land and quickly close around it.

Maybe plants can also perceive sound. A sort of clicking has been heard at the tips of roots. It probably arises when cell walls burst as they grow, and if other roots can hear this it might provide an explanation for a mystery. A single plant can have a million root ends, but they never get tangled. This suggests that they must somehow know each other's position. To me this seemed reminiscent of the way each individual in a flock of birds or a school of fish keeps track of their neighbours so they never crash. In any case, it is clear that plants use their roots rather like a brain, to gather information. This is, of course, just what Democritus believed.

The very possibility that roots could react to sound spurred fresh experiments. Under Mancuso's direction, a stereo system was installed at a vineyard. After five years, it was confirmed that vines close to the music fared better than others. The grapes matured earlier and had more colour and flavour, and as a bonus it seemed that the music confounded harmful insects, so the growers could dial down the pesticides. But the crucial factor for the grapevines wasn't the melody – it was the sound frequency. Certain intervals in the bass register seemed to support growth, while higher frequencies were inhibitive.

This theory was supported by a Japanese grocery company's discovery that mushrooms grew better to the sound of drums. Australian researchers picked up the thread – they found that wheat plants reacted to a tapping sound at 220 hertz and oriented their root tips in that direction. Could these organisms have been affected by vibrations? Judging by Mancuso's findings, plants could sense the Earth's electromagnetic fields, just like many animals can. Cress plants were happy to align their roots along them.

The questions only multiplied. Could plants' sensitivity be considered feelings of some sort? According to wine-growers, harvested wine was sometimes disturbed in a baffling way. It happened twice a year, first when the vines were blossoming and then during harvest time. On these occasions, young wines could grow cloudy for a few days, even though they were stored in barrels or bottles far from the harvest site. Was this a lingering, inherited rhythm of life, or an expression of sympathy? If plants had a type of consciousness, it wasn't unreasonable to inquire about their emotions.

These were the sorts of questions that had prompted us to wonder about everything laid out on the veranda table. If even wine and wheat could sense their surroundings, perhaps awareness is pretty much everywhere. Were we sitting in the midst of a network of plant and animal feelings? Was this, perhaps, a distinguishing feature of life?

The energetic children had been out of sight for a while as they played hide-and-seek, but it was getting near their bedtime. Their dads had adopted the summer tradition my sister and I had once kept, holidaying for a week together without their other halves. Since both the children and the dads would be sleeping in the cottage, a little rearranging ensued as my sister spent some time with the small ones.

Meanwhile, I took an evening jaunt around the property. What was really going on in all those sap-rich trunks, expansive branches and far-reaching root systems? Communication among both tree roots and ants was probably underway beneath my feet, and trees too were strengthened by their contacts. Together they would build forests that were like superorganisms.

Perhaps the same could be said of our cultures and societies. But plants have keener contact with the Earth, given that they literally live within it. And here Mancuso provided words of comfort in keeping with the time; he believed that future technology would transform plants into interpreters, giving humans information about air and soil quality, clouds of toxins or imminent earthquakes. Together, plants would be a 'Greenternet'.

They had, in any case, proven that they understand quite a bit about the world. They react to light, sonic vibrations and chemicals, and they are constantly in motion. I only perceive them to be standing still because my vision and brain interpret time differently. Plants simply live in a different rhythm of time, world of senses and system of communication, so I should hardly extrapolate my own ways of measuring and expression into a general norm.

Then again, everything alive has a related system of communication deep down. It exists on a cellular level, and it seems that the animal and plant kingdoms nearly meet within it. Animal cells are actually similar to the cells at the outermost edges of plants' roots.

This discovery was made by Darwin as he compared plant roots to the brains of earthworms. A major part of his research was on these very topics, botany and earthworms, although the latter was only discussed in whispers for a long time. It was bad enough that he thought our forefathers were apes – if the general public had found out that this family tree encompassed worms as well he would have been dismissed entirely. To add insult to injury, he even considered worms to have a certain amount of intelligence. As a result, his research about them was suppressed until the 1930s, when it was unearthed by the American inventor of the steel plough.

It wasn't news to anyone that earthworms work the soil. In ancient Egypt, Cleopatra forbade their export, for without worms the Nile Valley would never have been fertile enough to birth a prosperous civilisation. Industrious worms can, in just about a decade, turn over a layer of soil a decimetre thick, aerating it and restoring its nutrients.

But Darwin's interest had nothing to do with the earthworms' talent for farming. Rather, he was preoccupied by their similarities with plants. He had first noticed these when he tried to understand earthworm senses, testing their sight by placing lamps next to them and studying their hearing by playing instruments for them. Neither the bassoon nor the penny whistle provoked a reaction, nor was there any effect when he shouted at them. But something did happen when he put them in a pot of soil on the piano. The notes caused them to burrow beneath the surface. Like plants, then, they could sense vibrations in the earth. Did the changing tones carry a message? Worms themselves, it turns out, produce faint but regular sounds.

When it came to their sense of smell, they didn't seem to care about perfume or tobacco smoke. They did, however, react to the smell of their favourite foods, so Darwin was able to ascertain what they preferred to eat. They taste their food just as we do, and they appeared to find leaves of the wild cherry tree even more delicious than hazelnut leaves. Cabbage, carrot, celery and horseradish were other favourites, but they would hardly touch herbs like salvia, thyme and mint.

When Darwin summarised his comparisons, the similarities between worms and plants were striking. Since they both live in the ground, they have similar senses. Like plants, worms have a special sense of touch as well as photoreceptors instead of eyes. Like plants, worms can assess the chemistry of soil without taste buds or noses, and like plants they consist of multiple segments that can survive without one another. Even though worms aren't grazed like grasses, they're eaten by everything from badgers to birds and even fish sometimes. It's not enough, then, for earthworms to produce many young. They must also be able to lose part of their body without dying. In the name of science, these poor critters have been divided almost forty times before giving up.

Their most important parts are their fronts. Like the tips of roots, they are both strong enough to drill through soil and have the ability to release alarm pheromones. Worms release these when they're in uncomfortable spots they want to avoid in the future, but they also sometimes do so when they're speared on a fishing hook. On occasion, fish can sense this substance and are deterred from biting.

Naturally, though, worms are different from plants. A worm's red blood, for instance, is pumped around by five hearts, although they're really more like thickened blood vessels with muscular valves. It also has a brain, and while it's little more than an enlarged bundle of nerves it functions well enough to allow the worm to make decisions, orient itself and learn new things. Worms carefully select which leaves they want to eat, and when they transport food into their soil tunnels they try out different solutions to find the most effective way. Their smooth bodies have tiny brushes to grip the soil, and once they've rolled up a leaf and pulled it in, they plug the opening as protection against early birds.

In other words, those naked little living intestines turned out to be more complex than people expected. What's more, in the 20th century, it was found that the use of worms in folk medicine was justified. They can be used for pregnancy tests and contain a fever-reducing substance, so their chemistry is far from banal. But even so, their most interesting feature was their similarity with the root tips of plants, since it paved the way for a bigger question. How sharp are the lines between categories in biological taxonomy? After all, plants and animals have a common origin, and some organisms almost seem to have connections to both kingdoms.

A minor incident occurred as the children were being tucked in. A well-trafficked ant path was discovered near one of the bunk beds. Reactions were mixed. One of the children was fascinated and wanted to know all about ants, but the others wanted to get rid of the intruders on the spot. I sympathised with both teams. To start with, I could try to find one of the poison traps I had sent for. They should be somewhere in the carpentry shed.

This was where I had seen something else that might interest the children. One day, a yellow cushion had appeared on the decaying chopping block. At first I thought it was something the fox had brought over but, on closer inspection, it turned out to be a slime mould called scrambled egg slime. It wasn't hard to understand why this variety was sometimes called 'troll butter' in Swedish. The slime mould could move. It's not typical for fungi to move above ground, so some biologists elected to count slime mould in the animal kingdom. This upset other biologists, for according to them slime mould belonged to the plant kingdom, because it spread spores.

None of the bickering biologists was correct. Slime moulds are neither animals, plants nor fungi. They're more like amoebas, given that they are made up of a single cell. But even simple cells are capable of many things. They can recognise each other, communicate and remember, and if, like slime moulds, they lack cell walls they can unite into a single cell with many nuclei. As these nuclei stream back and forth, the entire cell clump moves, and as it does so it greedily dissolves all the bacteria it can find. Slime mould can eat. No wonder it confused systematists.

Now that I was looking for the troll butter, it had vanished from the chopping block – and an indolent clump of cells probably wasn't all that exciting for the children anyway. But for researchers, slime moulds have turned out to provide important information about the evolution of life. Their many nuclei illustrate what can happen when single-celled organisms begin to collaborate. When slime moulds were tested in mazes they showed evidence of a remarkable memory. Like ants, this memory manifested as scent trails they left behind, although the difference was that slime moulds absolutely did not want to return to their own trails. Quite the opposite – if they recognised their own scent, they took a different path to avoid places where they'd already eaten. Still, this provided researchers with a clue. An external memory like this was probably the first step on the path to an internal memory.

It could hardly be called primitive or poor, for our own external memories are the cornerstones of culture. I myself had very recently lost a family game of Memory, but I moved quite freely among such external memories as books. What slime moulds showed us was the very fact that external memory has played an important role in life, even when not in written form.

And what's more, it seemed to me that the troll butter said more than just something about the evolution of memory. It also proved a point about plants. Even an organism without eyes, a brain and a nervous system could orient itself in its surroundings, and remember.

Once I'd found an ant trap, I lingered outside the cottage for a moment. The honeysuckle had just begun to release its scent, and it billowed forth as luxuriously as the blossoms. Soon the evening moths would rustle and swarm among the nectary grottoes of the unfurled petals.

The scent gave an essence of a summer night and roused half-forgotten memories from past evenings when I had watched the flowers climb up crevices. After all, it wasn't only among ants and troll butter that scents could conceal memories. My sister still remembered the moment our mother showed her a butterfly orchid: 'Take a sniff!'

With that, the forest glade opened into a little world that lived on half a century later.

It was really quite astounding that incorporeal scents could reinvigorate something dead and gone. Marcel Proust, with the help of a sweet little sponge cake, unravelled the memories of an entire life. It's telling that the sense of smell fades in those with dementia, for while scents are as fleeting as the present moment, they can encircle events and places a person left behind long ago. They have guided us through life for millions of years, which is why we, like all mammals, have two nostrils to help us find the source of odours. This became less crucial once we began to draw maps with names for what we had seen, and at that point smells slipped into a more anonymous world of shadows.

But for deaf-blind Helen Keller, the world of smell was still expressive. Based on what she smelled, she could describe landscapes of meadows, barns and pine groves just as surely as if she had seen them. She also characterised people she met by their smells, more or less the way others recognise their friends by voice. She could tell if they had come from the garden or the kitchen, and she identified especially strong smells from those who had great vitality.

The sense of smell was equally important for Kaspar Hauser, who grew up in a dark cell. He could tell fruit trees apart by the subtle scent of their leaves. But for a long time he was helpless when faced with all the impressions that welled forth among humans.

Are the floating molecules of scents perhaps one of the most ancient expressions of life? Like the pheromones of the animal kingdom, smells provide an unadulterated essence of life, and they have been a crucial help for millions of years. Newborns find their mother's milk by smell, and bad odours warn of spoiled food. Even at a distance, smell can tell us things about others. Was that approaching creature a predator, potential prey or perhaps a partner? Hundreds of thousands of organic molecules surround each being, forming a unique signature.

When smells cross species lines they are interpreted more ambiguously. The aromatic scent of a coniferous forest is a deterrent for microbes, because it contains terpenes; ticks, moths and fleas hate lavender for the same reason. As for us, we appreciate the same fragrances as bees. That's why we borrow the scents of flowers, more or less as butterflies attract mates by smelling like roses. For thousands of years we have created perfumes out of petals, fruit rinds, seeds and leaves – yes, even from roots and bark. These ethereal essences can be combined just like tones in music, and in fact they are sorted into base notes, heart notes and top notes. A 19th-century perfumer made an entire scale of them, where D was violet, E was acacia, F was tuberose, G was orange blossom, A was freshly mown hay, B was southernwood and C was camphor. Other floral scents could create other scales, for in the world of smell there are just as many variations as there are in music.

The top notes reach the nose first and are first to fade. Heart notes run from jasmine and roses to the dried buds of cloves. Among the base notes, dried oakmoss can smell like the seashore or a rainy forest. But the classic base note is sandalwood. It's said to be calming and erotically arousing all at once, for tree essences are warm.

There are also animalistic base notes, such as ambergris, that mysterious essence that was once valued as highly as gold and slaves. It retains its scent for years and comes from deep within the sea. It can be found on beaches, but this fatty substance once enveloped the bones of cephalopods in the stomach of a sperm whale.

Faint or flourishing, mild or exciting – the scents of perfumes are gathered from all the varied places where life roils and streams. Like life, and like music, they slowly change and fade away, and still their quietly intense language has always followed us. In our very early days, our sense of smell was just a knot of tissue at the end of a nerve, but eventually it grew into a brain. Thus at one time the hemispheres of our brains were similar to buds on a smell-stalk, not unlike a sprouting flower. It's even been suggested that thoughts arose by way of our sensing odours. But thoughts certainly don't belong to the sphere of smell-sense. Their roots are in the ancient limbic system, the emotional centre of the brain, so scents are linked to emotions.

Just like feelings, they can also be difficult to describe. For how do you capture a scent? Roman poet Lucretius believed that the sense of smell could map out the shape of scent particles. A similar theory was suggested in the 1960s, in which the molecules of floral scents were said to be wedge-shaped, those of musky scents disc-shaped and those of camphor ball-shaped. But neither the shapes of scent molecules nor their chemical formulas have made it easier to render essences into words. Perfumers who can differentiate between thousands of smells find themselves at a loss when asked to describe them. Scents belong to a language that has not been tamed by any grammar. They are floating chemistry, spread on the breeze, on moisture and heat, companions of the present and the life of Earth itself.

Perhaps the emotional root of our sense of smell can also explain why so many things are expressed through flowers. Bouquets are used everywhere from birthday celebrations to graves, and there are even flower-language books to explain to couples in love exactly which emotion can be expressed by which species. This was a dodge by the Victorians, who preferred to use flowers as euphemisms instead of mentioning what they and bees did together. Anyhow, it illustrated how flowers have helped us with our expressions.

And it's not just about emotions. Although our eyes can see millions of shades of colour, we can only talk about the handful that have been given names – so here, too, flowers have come to our aid. The name 'rosy' comes from the rose, and the fruit called orange gave us the colour name; 'violet' comes from violets, and 'lavender' from the flower. An archaic word for another floral shade of purple, gridelin, comes from French *gris de lin*, or, 'the grey of flax'. With the advent of this word a sudden brightening of the palette also arrived, because earlier that shade of colour was called brown in some places.

In fact, colours and scents speak to us as powerfully as they do to bees, but talking about them is another story, and it's even more difficult to bring them to life. Written words have been carried forth on dead plants – on dried papyrus grass in Egypt, and on cut slices of beechwood in Scandinavia. Since then, billions of words have been transmitted on paper made from pulped trees. Together they must create something greater, as when bees make honey from blossoms, for the result is meant to convey an essence of life that will last into the future.

Was this perhaps the point at which literature intersects biology? After all, 'culture' does mean 'cultivation'. And ideas can certainly be cross-pollinated just like plants, and new branches grafted to them like fruit trees. Overcomplicated sentences can be weeded like flowerbeds to reach a more organic rhythm, and words can be transplanted into other languages or give rise to new hybrids. Many can branch off into flowery associations, and together they bring scent and shadow to a written world.

When I thought about it, literature and the art of gardening seemed to have many points of contact. No one can compete with nature, but we, too, can give time and care to the expressions that strive to spring forth. So it wasn't just the tradesmen I felt a kinship with. I also had an affinity for the gardener.

* * *

The days of holiday with my family were intense even in their timelessness, and afterwards it truly felt as if time had been packed into a seed loaded with growing power. Memories can, just like foliage, have a long reach, so they would linger even when everything they touched upon was gone.

The younger generations had continued their holidaying at the cottage while I was elsewhere. When I returned to meet the gardener, shadows were growing among blueberries in the forest. But there was also a bigger change afoot. Wells in the neighbourhood had begun to run dry during the summer drought, and almost all the birches on the property had died. Their leaves had fallen in August, never to return.

I recalled how the fate of Yggdrasil was described in the *Edda* and felt dejected. There, the snake Nidhogg had chewed its way ever deeper into its root, and the water in Mímer's well, which nourished the second root, slowly began to sour. At last only the third root remained, where the Norns spun, plied and cut the thread of life. They were worried when they saw the leaves yellowing on the branches, but the people of Midgard continued to live life as they had before, until the tree was felled by a storm and the water came rushing in. The gods had abandoned the World Tree to a fire that made the sky burn red.

As I walked around the house, I found that at least the birch at the corner had survived the drought, perhaps because its roots ran deeper. It must have sensed what had happened to the other birches on the property, since they were family, and trees can communicate. On the other side of the house stood the dry skeleton of its sister birch, a sad, eye-catching sight in front of the door. This was where I had hung a swinging birdhouse that first spring. The tree was always full of birds, and in June I had seen a sparrowhawk dive into the branches to take a blue tit as easily as if it were picking a piece of fruit. It was a painful sight, but the tree welcomed all birds.

And couldn't even the most luxurious blossom hint at the fact that death is part of life? In the spring scent of lilacs one ingredient, the organic compound indole, is created from putrefaction. The same double nature is present in the summer constellations of umbelliferous plants. On the one hand they include parsley, parsnip, cumin and chervil, which give us flavour and healing; on the other hand, poison hemlock: death in the pot. They're all part of the same family and can only be differentiated by features of the stem, leaf, fruit, root, flowering period and habitat. Between life and death there is a strange relation, like the right side and wrong side in a piece of knitting. Half of the species in a forest live on dead trees, and plants take nutrition through the process of decomposition that creates soil. Yet the constituent parts always remain. They are indications that the fertility of the Earth was born in an interplay between all living things.

When I had raked up the dead leaves on the ground and brought them to the compost pile it almost seemed to be the most active corner of the property. Down in the humus was a landscape as varied as the one I could see around me. Pollen mingled with ground-down bedrock, bacteria and countless other minuscule organisms. Somewhere in the forests of fungal mycelia were the love nests of spring-tails. Somewhere a beetle was finding its dinner and a millipede was eating a woodlouse. It was a teeming, often nameless life, for only a fraction of underground species are known. Yet all together, they're the ones who create the ground itself. The ancient Greeks considered the earth an element, but it is a fluid collaboration of water, air, particles and innumerable miniature lives.

I dug up a shovelful of compost and took a look at it. There were probably millions of bacteria there, a hundred thousand microscopic worms, and perhaps twenty thousand mites among various fungi and algae. These billions of voracious eaters would transform decaying material into food for foliage. They ate and drank as if at a giant banquet. Yeasts that could give us beer and wine, cheese and bread were now turning the sugar in leaf litter into alcohol, which bacteria gulped down so they could give acetic acid to others. It was combustion as strong as a flame on a candle, and it made everything go around like the world itself, so that the end of one story could be the beginning of another.

Dealing with the compost I got a tiny cut on my hand, so to be safe I went inside to clean it out. Bacteria are stealthy. Since their vast number of species have divided the world among them, the ones that nourish plants can be fatal to us.

So many remarkable things happen in the world of microbes. It's not just that different types of soil have their own specific array. All of them are waging a silent war against fungi, which defend themselves with a sort of antibiotic weapon. It seems that without knowing anything about the battle between bacteria and fungi, people in ancient Egypt covered wounds with a mash made of mould, so perhaps they had stumbled across its positive effects. When the connection became clear in the 20th century, it was still by sheer coincidence. Alexander Fleming forgot a dish of bacteria in his warm laboratory, and when he found it had been invaded by mould he understood the significance.

As a child I had been a battleground in the war between bacteria and fungi. One overzealous doctor wanted to cure my recurring infections once and for all by giving me a massive dose of antibiotics. He said it would kill all the bacteria, and I suppose in some ways it was a success. I suddenly became indescribably exhausted and allergic to just about everything. We had dogs at the time, and taking them on walks in the woods became torture because it seemed I could no longer tolerate plants or animals. But I didn't want to live in a sterile environment, so at last my mother took me to a naturopath, who made my allergies disappear. Perhaps the balance between fungi and bacteria was restored as I got stronger. I remember that the cure involved, among other things, pollen dragées.

Only later it was discovered that all the bacteria in a body together would weigh as much as a brain, and that they, in their own way, are just as important. Some types train the immune system, some bring out enzymes or vitamins and free nutrients; some keep foreign bacteria on the skin in check; some send neurotransmitters to the brain, suggesting that they may play a part in depression, autism and ADHD, among other things. Indeed, we humans harbour entire ecosystems of bacteria, and they're all necessary in the proper locations and proportions. Based solely on our billions of bacteria cells, we could almost be classified among them.

And it's not just that there are so many of them. They are full of incredible vigour. A bacterium that spent 250 million years inside a salt crystal came to life with a bit of moisture and soon began to multiply. Multiplication is child's play for bacteria. Since they divide every twenty minutes, they propagate as incessantly as they die. They can freely move towards nutritional sources, and they can sense changes in light, temperature, chemistry and magnetic fields. They can also exchange both molecules and DNA with one another, and, like ants, they can create vast networks. To boot, they're found in all living things on Earth, for there are reasons we can coexist. We have a common origin, in our oceanic past.

The early role of bacteria in evolution was discovered quite recently. In the 1960s, young biologist Lynn Margulis had a hypothesis about their importance, but at the time she was only considered a bizarre rebel, because she was going against the usual interpretation of evolutionary theory. That interpretation described evolution as a battle, but to her it seemed more about complementing than about competition. Perhaps natural selection could winnow, but it couldn't create something new. Cooperation could, however, create, because to add gave more than to take away. What's more, the survivors were most often those who had adapted to the world around them, and the world around them meant other kinds of life. To put it simply, everyone depended on everyone else. It wasn't as if the Earth housed eight million species for the sole purpose of benefiting a single one of them.

Margulis also found it misleading that zoologists illustrated evolution with animals. The animal kingdom was a recent arrival to life, so she looked back at the first cells of the primordial ocean. Even there, she saw signs of symbiosis. Her hypothesis was that bacteria had, at one point early on in life, made their way into other cells, where they became a sort of powerhouse. It seemed probable that this was how life was able to increase its diversity. In time, her theory would be corroborated by genetic studies, and she would come to be called one of the most prominent researchers of modern biology.

But it took some time for her to get anyone's attention. Having bacteria in the family tree was even worse than being descended from apes, and the article in which she presented her thesis was firmly rejected by fifteen scientific journals. Yet she continued her research undaunted.

She had noticed, for instance, that bacteria produce gases and suspected that this had affected the atmosphere. Since she was married to astrophysicist Carl Sagan, she knew that microscopic and astronomic perspectives can be reconciled. And she knew that her theories were shared by biochemist James Lovelock, who worked for NASA.

Lovelock had compared the Earth's atmosphere with those of Mars and Venus, and from the differences he had drawn an essential conclusion: the origin of our atmosphere is biological. It seems to have been created by the living organisms on Earth, which appear to regulate it still. The atmosphere – what we call the sky – is, like the soil, a creation of everything that has lived on Earth and is, therefore, unique.

In a sense, it had become easier to see the Earth from space than from the limited perspective afforded from the ground. The first astronauts were amazed at our shimmering, blue-and-green little pearl. They could see how rain and wind moved freely around it, for there were no borders, only transitional spaces, like between a mountain and a valley.

When Margulis and Lovelock combined their knowledge, the very smallest and the very biggest perspectives engaged one another – and that was telling. They discovered an interplay that reached from the cells to the atmosphere, with constant loops of feedback. It was like a woven fabric in which plants, animals, fungi and microorganisms were interdependent. Evolution was not a ladder moving towards something better, and the new was not more robust than the old. The very bacteria that were first on the scene would, without a doubt, out-survive everything else.

So what should this hypothesis be called? Lovelock liked to take long walks with his neighbour, author William Golding. Golding would later receive the Nobel Prize in Literature, but as a young man he had also studied the natural sciences. When Lovelock told him of the hypothesis that emerged from his collaboration with Margulis, Golding was enraptured. Why not name it after the old Earth goddess Gaia? Her Greek name was also Ge, as in 'geology' and 'geography'. The fact that Golding, as a novelist, was accustomed to a main character driving the action probably played a part here. In the Greek pantheon, Gaia was the goddess of the Earth, mothering and fertility, and since it's easy to assign some human traits to gods, this could promote understanding of the concept. Lovelock, a romanticist, accepted Golding's suggestion, and thus the Gaia Hypothesis got its name.

But Lynn Margulis was never satisfied with it. The goddess Gaia was a metaphor that led to faulty associations. The hypothesis had nothing to do with any single source of power – quite the opposite, it was about a flexible interplay between all the organisms on Earth, from bacteria to plants and animals. It was tempting to see some human traits behind whatever guided life on Earth, but this theory was about the reverse.

Margulis's scepticism turned out to be warranted. Because of its name, the Gaia Hypothesis came to be associated with occult, New Age fantasies. It was also interpreted as a feminine principle, so in the same spirit Margulis herself would be termed 'science's unruly earth mother', more or less like a modern Gaia. The central idea of the hypothesis – the link between all life on Earth – instead came to be tied to a different name. Aristotle, after all, had compared the connections on Earth to a joint household, and the Greek word for house brought us the term 'ecology'. It was first used by biologist Ernst Haeckel in the 19th century, and it began to be used more generally in conjunction with the burgeoning environmental movement of the 1960s.

As for Lovelock's and Margulis's theory, it came to enjoy support in most scientific circles with time, but the Gaia name was studiously avoided. Today it's more common to talk about geophysiology or 'Earth System Science', and the interactions of Earth are illustrated with a technical image: it's compared to interconnected computers. This was also the way Stefano Mancuso explained the linked intelligences of ants and plants. Big or small – it's all about networks.

When I was done with the compost, I sat down on the veranda and pondered all the different depictions of Earth's interplay. A net suggests both strength and vulnerability, since each individual loop is important. It's more than just a common thread, for in the soil threads become mycelia like those beneath the trees. And when I looked at the birch at the corner of the cottage, I thought that even the old World Tree was a valid symbol. Unlike goddesses, houses and computers, a tree is an actual living thing on Earth. The bark of the birch was wrinkled like elderly skin, and yet I knew that the buds of spring were lying in wait in its twigs. They all shared one trunk, even though the tree was a genetic mosaic with cascades of possibilities. It was only through its very diversity that the tree could live. Each part was perfect in its own way, for even if life is profligate, nothing is meaningless.

The family tree I had drawn for the little ones over the summer was meant to encompass our family. But where was the line that separated us from all else? In a much broader family tree, researchers by now had found the smallest common denominator for all life on Earth. It was a sprawling congregation that extended from microbes to plants to animals. This origin was called LUCA, for Last Universal Common Ancestor. Perhaps LUCA had arisen in multiple places, but it was, in any case, a primordial cell.

My eyes swept across the birch, the blue tits and the sparkling sound that was their backdrop. If LUCA had descendants in all living things, the cells in my body must be related to those in other beings, so apparently I had family members all over the property. Even though we looked nothing alike, we had our innermost parts in common.

It seemed inconceivable that cells could build such diverse organisms. But the secret, of course, lies in the genetic alphabet that constantly creates new worlds, more or less the way letters and books do. Some million genes inside me are proof of that. One of them contained the blueprints for my nervous system and is also found in insects and worms. And that's not all. I also share genes with trees and lilies.

Naturally, the property was home to only a selection of all life on Earth. I had got a broader overview in a museum of natural history that claimed to show the breadth of life in a segment. It started with the Big Bang, majestically projected onto the dome of the movie room, and as all the elements were formed, a sky teeming with stars unfurled. Zooming in on newborn Earth, volcanic eruptions flared under meteor showers. Then came the glowing red Earth, followed by the Earth black with lava and the Earth white with frost, before the green Earth began to flourish with life.

I walked on through halls of minerals, patterned with time or squeezed into gems. I saw the impressions of fossils among bones of cephalopods and skulls of ancestors. I passed caravans of animals, extinct or merely dead. When I reached an expressionlessly staring badger I hurried to a room where butterflies had been pinned up like vanished moments. Under them, beetles formed shimmering armies, and other displays glowed with the feathery suits of birds, preserved with arsenic. A famous herbarium sagged under the weight of its dried rarities.

But something was missing in these halls and exhibits. It was the life that constantly reached ahead, to nourish itself, to mate, communicate, hunt or flee, all in close contact with earth, water and sky, and with one another.

This was the eternal interplay that allowed everything to thrive, insatiable in its hunger for life, and for even more life. The walls people had erected between different species had two sides that belonged together. I knew now that the walls in the house where I would sleep were insulated with ants and bees, and the ceiling was a floor to birds even as the floor was a ceiling for foxes.

But one question remained. How was this living whole held together? I thought about how a myriad of dots can create figures in pointillistic art, just as the pixels on a screen do. The more dots, the sharper the picture, so each dot contributes to shades and details.

Gathering them into one narrative was impossible, and in a story everything takes place from one viewpoint. That's not the case in the teeming masses of life. But in a painting class I had learned a method to deepen perspective by using the golden ratio that came out of the Renaissance. Remarkably enough, it seems to be mirrored in nature. Author Peter Nilson found the proportions of the golden ratio reflected in the structure of snails, pinecones and sunflowers, so perhaps nature and art obey similar laws.

As an astronomer, Nilson even found points of contact between the shape of the universe and that of music. The vibration-like movements of the first atoms continued through the cosmos and are the cause of so-called 'flicker noise' in computers. This was discovered not only in distant star systems, but also in the noise of Earth's waterways and winds, in natural disasters and in the variations of stock markets.

There are related sound patterns even among living creatures. When a gibbon's song is played at double speed it sounds like birdsong; when played slower it sounds like whale song. When the sound waves are written out they all have the same pattern, rather like the way a twig resembles a tree. The differences lie in scale and tempo.

And tempo compensates for some of our other differences as well. In a single second, a bee can comprehend movements a hundred times faster than what I can see. Smaller animals with faster metabolisms take in more of the world than bigger ones do. The hearts of songbirds and mice beat 600 times per minute, as if they were leaves fluttering in the breeze. The tempo of a whale's heart is a hundred times slower, so lifetime heartbeats can number about the same in creatures large and small.

Could these be seen as time signatures? If insects move in sixteenths and mammals in fourths, the plodding gait of a badger would be a whole note. Every creature moving through a flowing piece of music with endless variations. Beneath them all runs a common chord in the magnetic field at the core of the Earth. It fluctuates between eight and sixteen times per second. The same rhythm, in my brain, formed a state of quiet. What a mind-boggling thought. Could we be tuned to the Earth?

Then I remembered NASA's recording of the Earth's electromagnetic vibrations. Transformed into sound, it was a roaring harmony with no beginning or end. Could all life on the planet be part of that roar? Biochemist Jesper Hoffmeyer called the Earth's entire sphere of communication the semiosphere, and as a biological grammar it encompassed millions of modes of expression. It was scents, colours and shapes; chemical signals, touches and movements; waves of all sorts and electrical fields – in short, signs of life. It was the modulations of the blackbird, the fifth of the great tit, the buzz of insect wings. It was whale song and the various sounds of fish among the drumming of molluscs. It was the howls of mating calls, the burbling snorts of mother foxes, the grunting of badgers, the ultrasonic songs of field mice and the faint tones of earthworms.

At the core of every being, too, lay silent, vibrating chords of genes. They have been compared to music played on an instrument with four notes. One gene could contain a hundred chords, and in tandem with others could turn variations on an ancient theme into continual new tones. It never ceased, for life is an ever-unfinished symphony.

The air had that bright, gentle, late-summer feeling. It was getting to be time for the migratory birds to take off. They were surely storing their meagre provisions before watching the landscape change beneath them.

I no longer dreamed of following them. The saturated air they were borne by was here as well. It held thousands of scents, vibrations of wingbeats and millions of bygone breaths. There were even scattered parts of life in the air, for the molecules had traces of a hundred different algae, forty thousand fungi spores, and the pollen of ten thousand plants. Among them, too, hovered microscopic particles of salt, ash, mud and even topaz, as if the entire world wished to gather in the limitless air. I myself contributed to it, for each hour I shed a million tiny particles of skin, many with an invisible passenger or two.

My senses could capture none of this. Even the nerve cells that built a world in my brain were unknown to me. There were as many of them as there were stars in the Milky Way, so everything I perceived was encompassed in their immeasurable branches. In the same instant I had a thought it was re-created there, and when my attention was captured by something else it lived on in the background, just as everything on the property did when I looked away.

Yet I could only perceive a fraction of everything around me. In my mind I measured my own abilities against the keen senses of other creatures. Foxes can hear an earthworm's bristles move against the grass, and plants' roaming roots can perceive the faint chemistry of the soil. Eels can pick up a thimble's worth of an essence that has been sprayed in a lake, while dolphins use echolocation to understand the nature of something a hundred metres away. Compasses, weather radar and GPS fit in the brains of migratory birds, and a male mosquito can smell a female at a distance of several kilometres. Ants, for their part, create the infrastructure of a whole society using scents.

And think of what bees can perceive! Their extensive internal maps contain an entry for each flower, plus the time it blossoms and the flight time to reach it. They can see ultraviolet shifts in flowers just as birds do in tree leaves, so they can fly in and out among them.

What if all of these specialised senses could be combined? What would they show? Are we in fact part of some image or music? It could be created in a stream that took its shape from what it filled, and perhaps this was how life had found a centre in the tiniest mite and blade.

Was I simply loving life through all the many shapes it came dressed in? The orpine was still blooming on the ground and I suddenly remembered that it was used for love magic. One stalk was bent, and I picked it to put among some twigs in a jug on the veranda table. Immediately a late-summer bumblebee came by to find some nectar. Wingbeats and rustling leaves swept me up in the sounds of life. Then a branch of the birch brushed the western wall, and I looked up to find a squirrel had just settled there. Judging by its skinny tail, it was one of a new generation who would soon take over the property. It regarded me sleepily, then closed its eyes for a moment. Then it opened them again to look me right in the eye. The Earth had a thousand species I had never seen and a thousand languages I had never mastered, but it also offered wordless encounters such as this one. I was happy.

Ratatosk, I thought. Please stay. We need to look after the tree.

Author's Note

THE WORLD IS TEEMING with information that often seems more like a fairy tale than any work of fiction could. In a book with thousands of facts from hundreds of sources, all are important. Had I cited them in the text it would have been an academic dissertation rather than a literary essay. Instead, the sources of all referred facts are gathered here, with many thanks to the countless researchers behind them.

Bibliography

Translator's note: original titles have been added for those works the author read in translation.

Introduction – Into Nature

Barnes, Jonathan, *Aristotle: A Very Short Introduction*, Oxford 2000

Burton, Nina, Gutenberggalaxens nova. En essäberättelse om Erasmus av Rotterdam, humanismen och 1500-talets medierevolution, Stockholm 2016

Farrington, Benjamin, Grekisk vetenskap. Från Thales till Ptolemaios, övers. Lennart Edberg, Stockholm 1965 (Original title: *Greek Science: Its Meaning for Us*)

1 – The Blue Roof

Ackerman, Jennifer, Bevingad intelligens. I huvudet på en fågel, övers. Shu-Chin Hysing, Stockholm 2018 (Original title: *The Genius of Birds*)

Alderton, David, *Animal Grief: How Animals Mourn*, Poundbury 2011

Bach, Richard, Måsen: berättelsen om Jonathan Livingstone Seagull, övers. Tove Bouveng, Stockholm 1973 (Original title: *Jonathan Livingston Seagull*)

Barnes, Simon, *The Meaning of Birds*, London 2016

Bastock, Margaret, Uppvaktning i djurvärlden. En bok om parningsspel och könsurval, övers. Sverre Sjölander, Stockholm 1967 (Original title: *Courtship: A Zoological Study*)

Bright, Michael, Intelligens bland djuren, övers. Roland Staav, Stockholm 2000 – Djurens hemliga liv, övers. Roland Staav, Stockholm 2002 (Original title: *Intelligence in Animals*)

Burton, Nina, Den hundrade poeten. Tendenser i fem decenniers poesi, Stockholm 1988

Caras, Roger, Djurens privatliv, övers. Bo och Gunnel Petersson, Stockholm 1978 (Original title: *The Private Lives of Animals*)

Chaline, Eric, Femtio djur som ändrat historiens gång, övers. Hans Dalén, Stockholm 2016 (Original title: *Fifty Animals that Changed the Course of History*)

Edberg, Rolf, Spillran av ett moln, Stockholm 1966

Fridell, Staffan & Svanberg, Ingvar, Däggdjur i svensk folklig tradition, Stockholm 2007

Graebner, Karl-Erich, Livet i himmel, på jord, i vatten, övers. Roland Adlerberth, Stockholm 1975 (Original title: *Natur – Reich der tausend Wunder*)

——Naturen – livets oändliga mångfald, övers. Roland Adlerberth, Stockholm 1974 (Original title: *Natur – Reich der tausend Wunder*)

Gorman, Gerard, *Woodpeckers*, London 2018

Griffin, Donald R., *Animal Minds*, Chicago 1992

Hagberg, Knut, Svenskt djurliv i mark och hävd, Stockholm 1950

Haupt, Lyanda Lynn, *Mozart's Starling*, New York 2017

Ingelf, Jarl, Sjukvård i djurvärlden, Stockholm 2002

Isaacson, Walter, Leonardo da Vinci, övers. Margareta Eklöf, Stockholm 2018 (Original title: *Leonardo da Vinci*)

King, Doreen, *Squirrels in Your Garden*, London 1997

Lagerlöf, Selma, Nils Holgerssons underbara resa genom Sverige, Stockholm 1907

Leroi, A.M., *The Lagoon: How Aristotle Invented Science*, London 2015

Linsenmair, Karl-Eduard, Varför sjunger fåglarna? Fågelsångens former och funktioner, Stockholm 1972

Lorenz, Konrad, I samspråk med djuren, övers. Gemma Snellman, Stockholm 1967 (Original title: *Er redete mit dem Vieh, den Vögeln und den Fischen*; English title: *King Solomon's Ring*)

——Grågåsens år, övers. Håkan Hallander, Stockholm 1980 (Original title: *Das Jahr der Graugans*; English title: *The Year of the Greylag Goose*)

Marend, Mart, Vingkraft, Klintehamn 2012

Meijer, Eva, Djurens språk. Det hemliga samtalet i naturens värld, övers. Johanna Hedenberg, Stockholm 2019 (Original title: *Dierentalen*; English title: *Animal Languages: The Secret Conversations of the Living World*)

Milne, Lorus J. och Margery, Människans och djurens sinnen, övers. Svante Arvidsson, Stockholm 1965 (Original title: *The Senses of Animals and Men*)

Nilson, Peter, Stjärnvägar. En bok om kosmos, Stockholm 1996

Rådbo, Marie, Ögon känsliga för stjärnor. En bok om rymden, Stockholm 2008

Robbins, Jim, *The Wonder of Birds*, London 2018

Rosen von, Björn, Samtal med en nötväcka, Stockholm 1993

Rosenberg, Erik, Fåglar i Sverige, Stockholm 1967

Safina, Carl, *Beyond Words: What Animals Think and Feel*, New York 2015

Sax, Boria, *Crow*, London 2017

Signaler i djurvärlden, red. Dietrich Burkhard, Wolfgang Schleidt, Helmut Altner, övers. Sverre Sjölander, Stockholm 1969 (Original title: *Signale in der Tierwelt*; English title: *Signals in the Animal World*)

Taylor, Marianne, *401 Amazing Animal Facts*, London 2010

Tinbergen, Niko, Beteenden i djurvärlden, övers. Inga Ulvönäs, Stockholm 1969 (Original title: *Animal Behaviour*)

Ulfstrand, Staffan, Flugsnapparnas vita fläckar. Forskningsnytt från djurens liv i svensk natur, Stockholm 2000 – Fågelgrannar, med Sven-Olof Ahlgren, Stockholm 2015

Wallin, Nils L., *Biomusicology. Neurophysiological, Neuropsychological and Evolutionary Perspectives on the Origins and Purposes of Music*, New York 1992

Watson, Lyall, *Lifetide*, London 1979

——*Supernature II*, London 1986

Wickler, Wolfgang, Häcka, löpa, leka. Om parbildning och fortplantning i djurens värld, övers. Anders Byttner, Stockholm 1973 (Original title: *Sind wir sunder: naturgesetze der Ehe*; English title: *The Sexual Code: The Social Behaviour of Animals and Men*)

Wills, Simon, *A History of Birds*, Barnsley 2017

Wohlleben, Peter, Djurens gåtfulla liv, övers. Jim Jakobsson, Stockholm 2017 (Original title: *Das Seelenleben der Tiere*; English title: *The Inner Life of Animals: Surprising Observations of a Hidden World*)

Zänkert, Adolf, Varthän – Varför. En bok om djurens vandringar, övers. Birger Bohlin, Malmö 1959 (Original title: *Das grosse Wandern*)

ARTICLES:

Birds have primate-like number of neurons in the forebrain, *Proceedings of the National Academy of Sciences* 13 June 2016

Bounter, David och Shah, Shailee, A noble vision of gulls, Summer 2016 issue of *Living Bird Magazine*

Burton, Nina, Den sagolika verklighetens genre, De Nios litterära kalender 2007

Denbaum, Philip, Kråkor, DN 8 feb 2018

Ekstrand, Lena, Därför är kråkfåglar så smarta, GP 18 dec 2016

Snaprud, Per, Så hittar fåglarna, DN 11 maj 2002

Svahn, Clas, 2,9 miljarder fåglar har försvunnit i Nordamerika på 50 år, DN 19 sept 2018

Symposium för Kungl. Fysiografiska sällskapet 14 september 2017 på Palaestra, Lund 2017, The Thinking Animal – are other animals intelligent?

Bugnyar, Thomas, Testing bird brains: raven politics

Emery, Nathan, Bird brains make brainy birds

Roth, Gerhard, What makes an intelligent brain intelligent?

http://classics.mit.edu/Aristotle/history_anim.mb.txt

https://www.natursidan.se/nyheter/talgoxar-som-attackerar-smafaglar-utspritt-fenomen-som-dokumenterats-länge

https://www.svt.se/nyheter/lokalt/skane/talgoxen-utmanar-schimpansen

https://fof.se/tidning/2015/6/artikel/var-smarta-smafagel

https://djurfabriken.se/kycklingfabriken

2 – Wingbeats at the Door

Ackerman, Jennifer, Bevingad intelligens. I huvudet på en fågel, övers. Shu-Chin Hysing, Stockholm 2018 (Original title: *The Genius of Birds*)

Bergengren, Göran, Meningen med bin, Stockholm 2018

Boston, David H., *Beehive Paintings from Slovenia*, London 1984

Bright, Michael, Intelligens bland djuren, övers. Roland Staav, Stockholm 2000 (Original title: *Intelligence in Animals*)

——Djurens hemliga liv, övers. Roland Staav, Stockholm 2002 (Original title: *The Secret Life of Animals*)

Caras, Roger, Djurens privatliv, övers. Bo och Gunnel Petersson, Stockholm 1978 (Original title: *The Private Lives of Animals*)

Carson, Rachel, Tyst vår, övers. Roland Adlerberth, Lund 1979 (Original title: *Silent Spring*)

Casta, Stefan & Faberger, Maj, Humlans blomsterbok, 1993/2015

Chaline, Eric, Femtio djur som ändrat historiens gång, övers. Hans Dalén, Stockholm 2016 (Original title: *Fifty Animals that Changed the Course of History*)

Comont, Richard, *Bumblebees*, London 2017

Dröscher, Vitus B., Hur djuren upplever världen, övers. Roland Adlerberth, Stockholm 1969. (Original title: *Klug vie die Schlangen, die Erforschung der Tierseele*; English title: *The Mysterious Senses of Animals*)

Galen i insekter. En berättelse om småkrypens magiska värld, övers. Helena Sjöstrand Svenn & Gösta Svenn, Stockholm 2016 (Original title: *A Buzz in the Meadow: The Natural History of a French Farm*)

——Den stora humleresan, övers. Helena Sjöstrand Svenn & Gösta Svenn, Stockholm 2018 (*Original title: Bee Quest*)

Goulson, Dave, Galen i humlor. En berättelse om små men viktiga varelser, övers. Helena Sjöstrand Svenn & Gösta Svenn, Stockholm 2015 (Original title: *A Sting in the Tale: My Adventures with Bumblebees*)

Graebner, Karl-Erich, Naturen – livets oändliga mångfald, övers. Roland Adlerberth, Stockholm 1974 – Livet i himmel, på jord, i vatten, övers. Roland Adlerberth, Stockholm 1975 (Original title: *Natur – Reich der tausend Wunder*)

Griffin, Donald R., *Animal Minds*, Chicago 1992

Hanson, Thor, *Buzz: The Nature and Necessity of Bees*, N.Y.
    & London 2018

Hansson, Åke, Biet och bisamhället, i Landskap för människor
    och bin, Stockholm 1981

Klinting, Lars, Första insektsboken, Stockholm 1991

Lindroth, Carl H., Myran Emma, Stockholm 1948

——Från insekternas värld, Stockholm 1963

Lloyd, Christopher, *The Story of the World in 100 Species*,
    London 2016

Meijer, Eva, Djurens språk. Det hemliga samtalet i naturens
    värld, övers. Johanna Hedenberg, Stockholm 2019
    (Original title: *Dierentalen*; English title: *Animal
    Languages: The Secret Conversations of the Living
    World*)

Milne, Lorus J. och Margery, Människans och djurens sinnen,
    övers. Svante Arvidsson, Stockholm 1965 (Original title:
    The Senses of Animals and Men)

Möller, Lotte, Bin och människor. Om bin och biskötare i
    religion, revolution och evolution samt många andra
    bisaker, Stockholm 2019

Mossberg, Bo & Cederberg, Björn, Humlor i Sverige. 40 arter
    att älska och förundras över, Stockholm 2012

Munz, Tania, *The Dancing Bees. Karl von Frisch and the
    Discovery of the Honeybee Language*, Chicago 2016

Nielsen, Anker, Insekternas sinnesorgan, övers. Steffen
    Arnmark, Stockholm 1969 (Original title: *Insekternes
    sanseverden*)

Russell, Peter, *The Brain Book*, London 1979

Safina, Carl, *Beyond Words: What Animals Think and Feel*,
    New York 2015

Signaler i djurvärlden, red. Dietrich Burkhard, Wolfgang
    Schleidt, Helmut Altner, övers. Sverre Sjölander, Stockholm
    1969 (Original title: *Signale in der Tierwelt*; English title:
    Signals in the Animal World)

Sverdrup-Thygeson, Anne, Insekternas planet. Om småkrypen
vi inte kan leva utan, övers. Helena Sjöstrand Svenn &
Gösta Svenn, Stockholm 2018 (Original title: *Insektenes
planet*; English title: *Extraordinary Insects*)

Thomas, Lewis, Cellens liv, övers. Karl Erik Lorentz,
fackgranskning Bo Holmberg, Stockholm 1976
(Original title: *The Lives of a Cell: Notes of a Biology
Watcher*)

Tinbergen, Niko, Beteenden i djurvärlden, övers. Inga
Ulvönäs, Stockholm 1969 (Original title: *Animal
Behaviour*)

Watson, Lyall, *Supernature II*, London 1986

Wohlleben, Peter, Djurens gåtfulla liv, Stockholm 2017
(Original title: *Das Seelenleben der Tiere*; English title: *The
Inner Life of Animals: Surprising Observations of a Hidden
World*)

ARTICLES:

Jones, Evelyn, Därför kan vi inte leva utan insekterna, DN 16
mars 2019

Nordström, Andreas, Kärleken till humlan hänger på håret,
Expr. 10 mars 2011

Ottosson, Mats, Lycklig av bin, Sveriges Natur 1/05

Pejrud, Nils, Humlor och blommor – en elektrisk
kärlekshistoria, svt.se/nyheter/vetenskap 21 feb2013

Undseth, Michelle TT, Insekter har medvetande, svt vetenskap
18 april 2016

Aktuellt i korthet. Särbegåvad. Att bin kan räkna … Sveriges
Natur 4/2018

Humlan känner igen ditt ansikte, Allt om vetenskap 17 aug
2007

Studie visar att insekter har ett medvetande, DN 19 april
2016, TT

Humlor – smartare än du tror, AB 24 feb 2017, TT

Symposium för Kungl. Fysiografiska sällskapet 14 september
2017 på Palaestra, Lund 2017, The Thinking Animal – are
other animals intelligent?

Chittka, Lars, Are insects intelligent?

https://natgeo.se/djur/insekter/bin-kan-ocksa-bli-ledsna

https://svenskhonungsforadling.se/honung/honungsskolan

https://www.biodlarna.se/bin-och-biodling/biodlingens-
produkter/honung

https://meinhoney.com/news/the-researchers-found-that-a-
honeybee-has-the-same-amount-of-hairs-as-a-squirrel-3-
million

https://www.bumblebeeconservation.org/bee-faqs/bumblebee-
predators/

https://tv.nrk.no/serie/insekter-og-musikk

3 – The Ants on the Wall

Bright, Michael, Intelligens bland djuren, övers. Roland
Staav, Stockholm 2000 (Original title: *Intelligence in
Animals*)

——Djurens hemliga liv, övers. Roland Staav, Stockholm 2002
(Original title: *The Secret Life of Animals*)

Burton, Nina, Det splittrade alfabetet. Tankar om tecken och
tystnad mellan naturvetenskap, teknik och poesi,
Stockholm 1998

——Det som muser viskat. Sju frågor och hundra svar om
skapande och kreativitet, Stockholm 2002

Caras, Roger, Djurens privatliv, övers. Bo och Gunnel
Petersson, Stockholm 1978 (Original title: *The Private
Lives of Animals*)

Dröscher, Vitus B., Hur djuren upplever världen, övers.
Roland Adlerberth, Stockholm 1969 (Original title: *Klug
vie die Schlangen, die Erforschung der Tierseele*; English
title: *The Mysterious Senses of Animals*)

Goulson, Dave, Galen i insekter. En berättelse om
småkrypens magiska värld, övers. Helena Sjöstrand Svenn
& Gösta Svenn, Stockholm 2016 (Original title: *A
Buzz in the Meadow: The Natural History of a French
Farm*)

Graebner, Karl-Erich, Naturen – livets oändliga mångfald,
övers. Roland Adlerberth, Stockholm 1974 – Livet i
himmel, på jord, i vatten, övers. Roland Adlerberth,
Stockholm 1975 (Original title: *Natur – Reich der tausend
Wunder*)

Griffin, Donald R., *Animal Minds*, Chicago 1992

Hölldobler, Bert & Wilson, Edward, *The Superorganism: The
Beauty, Elegance, and Strangeness of Insect Societies*, New
York & London 2009

Ingelf, Jarl, Sjukvård i djurvärlden, Stockholm 2002

Johnson, Steven, *Emergence: The Connected Lives of Ants,
Brains, Cities, and Software*, New York 2001 & 2004

Lindroth, Carl H., Från insekternas värld, Stockholm 1962

Lindroth, Carl H. & Nilsson, Lennart, Myror, Stockholm
1959

Maeterlinck, Maurice, Bikupan, Stockholm 1922 – Myrornas
liv, övers. Hugo Hultenberg, Stockholm 1931(Original
title: *La Vie des Fourmis*; English title: *The Life of the
Ant*)

Martinson, Harry, Vinden på marken, Stockholm 1964

Milne, Lorus J. och Margery, Människans och djurens sinnen,
övers. Svante Arvidsson, Stockholm 1965 (Original title:
The Senses of Animals and Men)

Nielsen, Anker, Insekternas sinnesorgan, övers. Steffen
Arnmark, Stockholm 1969 (Original title: *Insekternes
sanseverden*)

Russell, Peter, *The Brain Book*, London 1979

Safina, Carl, *Beyond Words: What Animals Think and Feel*,
New York 2015

Sverdrup-Thygeson, Anne, Insekternas planet. Om småkrypen vi inte kan leva utan, övers. Helena Sjöstrand Svenn & Gösta Svenn (Original title: *Insektenes planet*; English title: *Extraordinary Insects*)

Taylor, Marianne, *401 Amazing Animal Facts*, London 2010

Thomas, Lewis, Cellens liv, övers. Karl Erik Lorentz, Stockholm 1976 (Original title: *The Lives of a Cell: Notes of a Biology Watcher*)

Wilson, E. O, *Anthill*, New York 2010

——*On Human Nature*, Harvard 1978

Wohlleben, Peter, Naturens dolda nätverk, övers. Jim Jakobsson, Stockholm 2018 (Original title: *Das geheime Netzwerk der Natur*)

ARTICLES:

Exploderande myror, SvD 9 juli 2018

Johansson, Roland, Vägbygget som inte behöver planeras, SvD 9 feb 2019

Myror kan räkna, Allt om vetenskap nr 6-2011

Rosengren, Izabella, Kyssens korta historia, forskning.se 2017/02/14

Thyr, Håkan, Myror mäter med pi, Ny Teknik 2000

Wallerius, Anders, Prat med myror blir möjligt, Ny Teknik 20-08

4 – A Veranda with a Sea View

Ackerman, Diane, *The Human Age: The World Shaped by Us*, New York 2014

Beerling, David, *The Emerald Planet*, New York 2007

Black, Maggie, *Water: Life Force*, Toronto 2004

Bright, Michael, Intelligens bland djuren, övers. Roland Staav, Stockholm 2000 – Djurens hemliga liv, Stockholm 2002 (Original title: *Intelligence in Animals*)

Burton, Nina, Flodernas bok. Ett äventyr genom livet, tiden och tre europeiska flöden, Stockholm 2012

Capra, Fritjof, *The Web of Life*, London 1997

Caras, Roger, Djurens privatliv, övers. Bo och Gunnel Petersson, Stockholm 1978 (Original title: *The Private Lives of Animals*)

Carson, Rachel L., Havet, övers. Hans Pettersson, Stockholm 1951 (Original title: *The Sea Around Us*)

Chaline, Eric, Femtio djur som ändrat historiens gång, övers. Hans Dalén, Stockholm 2016 (Original title: *Fifty Animals that Changed the Course of History*)

Davis, K. S. & Day, J. A., Vatten, vetenskapens spegel, övers. Leif Björk, Stockholm 1961 (Original title: *Water: The Mirror of Science*)

Day, Trevor, *Sardine*, London 2018

Dröscher, Vitus B., Hur djuren upplever världen, övers. Roland Adlerberth, Stockholm 1969 (Original title: *Klug wie die Schlangen, die Erforschung der Tierseele*; English title: *The Mysterious Senses of Animals*)

Edberg, Rolf, Droppar av vatten, droppar av liv, Höganäs 1984
——Årsbarn med Plejaderna, Stockholm 1987

Ellervik, Ulf, Ursprung. Berättelser om livets början, och dess framtid, Stockholm 2016

Evans, L. O., Jordens historia och geologi, övers. Marcel Cohen, Stockholm 1972 (Original title: *The Earth*)

Graebner, Karl-Erich, Livet i himmel, på jord, i vatten, övers. Roland Adlerberth, Stockholm 1975 Original title: *Natur – Reich der tausend Wunder*)
——Naturen – livets oändliga mångfald, övers. Roland Adlerberth, Stockholm 1974 (Original title: *Natur – Reich der tausend Wunder*)

Harari, Yuval Noah, *Sapiens*. En kort historik över mänskligheten, övers. Joachim Retzlaff, Stockholm 2015 (English title: *Sapiens: A Brief History of Humankind*)

——Homo Deus. En kort historik över morgondagen, övers.
   Joachim Retzlaff, Stockholm 2017 (English title: *Homo
   Deus: A Brief History of Tomorrow*)

Henderson, Caspar, *The Book of Barely Imagined Beings: A
   21st Century Bestiary*, Chicago 2013

Isaacson, Walter, Leonardo da Vinci, övers. Margareta Eklöf,
   Stockholm 2018 (Original title: *Leonardo da Vinci*)

Kallenberg, Lena & Falk, Bisse, Urtidsboken. Från jordens
   födelse tilldinosauriernas undergång, Stockholm 1996

Kuberski, Philip, *Chaosmos*, New York 1994

Lloyd, Christopher, *The Story of the World in 100 Species*,
   London 2016

Meijer, Eva, Djurens språk. Det hemliga samtalet i naturens
   värld, övers. Johanna Hedenberg, Stockholm 2019
   (Original title: *Dierentalen*; English title: *Animal
   Languages: The Secret Conversations of the Living World*)

Melville, Herman, Moby Dick eller Den vita valen, övers.
   Hugo Hultenberg, Stockholm 2016 (Original title: *Moby-
   Dick*; or, *The Whale*)

Meulengracht-Madsen, Jens, Fiskarnas beteende, Stockholm
   1969

Milne, Lorus J. och Margery, Människan och djurens sinnen,
   övers. Svante Arvidsson, Stockholm 1965 (*Original title:
   The Senses of Animals and Men*)

Nicol, Stephen, *The Curious Life of Krill: A Conservation
   Story from the Bottom of the World*, Washington 2018

Nilson, Peter, Stjärnvägar, Stockholm 1991

Philbrick, Nathaniel, I hjärtat av havet. Den tragiska
   berättelsen om valfångsfartyget Essex, övers. Hans
   Berggren, Stockholm 2001 (Original title: *In the Heart of
   the Sea*)

Russell, Peter, *The Brain Book*, London 1979

Safina, Carl, *Beyond Words: What Animals Think and Feel*,
   New York 2015

Sagan, Carl, Lustgårdens drakar. Den mänskliga intelligensens utveckling, övers. Carl G. Liungman, Stockholm 1979 (Original title: *The Dragons of Eden: Speculations on the Evolution of Human Intelligence*)

Signaler i djurvärlden, red. Dietrich Burkhard m.fl., övers. Sverre Sjölander, Stockholm 1969 (Original title: *Signale in der Tierwelt*; English title: *Signals in the Animal World*)

Sörlin, Sverker, Antropocen. En essä om människans tidsålder, Stockholm 2017

Teschke, Holger, Sill. Ett porträtt, övers. Joachim Retzlaff, Stockholm 2018 (Original title: *Heringe*)

Thomas, Lewis, Cellens liv, övers. Karl Erik Lorentz, fackgranskning Bo Holmberg, Stockholm 1976 (Original title: *The Lives of a Cell: Notes of a Biology Watcher*)

Tinbergen, Niko, Beteenden i djurvärlden, övers. Inga Ulvönäs, Stockholm 1969 (Original title: *Animal Behaviour*)

Waal de, Frans, *Are We Smart Enough to Know How Smart Animals Are?* New York 2016

Watson, Lyall, *Lifetide*, London 1979

——*Heaven's Breath: A Natural History of the Wind*, London 1984

——*The Water Planet: A Celebration of the Wonder of Water*, New York 1988

Wege, Karla, Väder, övers. Thomas Grundberg, Stockholm 1993 (Original title: *Wetter*)

Welland, Michael, *Sand: The Never-ending Story*, Berkeley 2009

Wilson, E. O., *On Human Nature*, Cambridge, Massachusetts 1978 & 2004

——*Half-Earth: Our Planet's Fight for Life*, New York 2016

Wohlleben, Peter, Djurens gåtfulla liv, övers. Jim Jakobsson, Stockholm 2017 (Original title: *Das seelenleben der tiere*; English title: *The Inner Life of Animals: Surprising Observations of a Hidden World*)

Zänkert, Adolf, Varthän – Varför. En bok om djurens
    vandringar, övers. Birger Bohlin, Malmö 1959 (Original
    title: *Das grosse Wandern*)

ARTICLES:

Backman, Maria, Tyst, vi störs! Sveriges Natur 5/2017 –
    Känner med vän, Sveriges Natur 1/2020
Bertilsson, Cecilia, Utan öron, inga ljud, Sveriges Natur
    2/2004
Britton, Sven, Cellerna dör för nästan, DN 28 mars 1995
Brusewitz, Martin, Alger, hajp eller hopp ? Sveriges Natur
    2/18
Djuret som lurar evolutionen, Forskning och Framsteg 20
    april 2019
Djur i rymden, DN 6 feb 2013
Högselius, Per, Spår i sand berättar om jordens historia, SvD
    14 april 2011
Kerpner, Joachim, Vattnet på väg att ta slut i 17 länder, AB 7
    aug 2019
Livet uppstod i en vattenpöl, Illustrerad vetenskap 20 april
    2019
Schjærff Engelbrecht, Nønne/TT, 33 storstäder hotas av
    vattenbrist, SvD 7 aug 2019
Thorman, Staffan, Att leva i vatten, ur utställningskatalogen
    Vatten. Myt. Konst. Teknik. Vetenskap, Lövstabruk 1991
Symposium för Kungl. Fysiografiska sällskapet 14 september
    2017 på Palaestra i Lund 2017, The Thinking Animal – are
    other animals intelligent?
  Mather, Jennifer, Mind in the water
https://blueplanetsociety.org/2015/03/the-importance-of-
    plankton/
https://www.forskning.se/2017/09/18/fiskhonor-gillar-
    hanarsom-sjunger/
https://octopusworlds.com/octopus-intelligence

https://svt.se/nyheter/inrikes/odlad-lax-full-av-forbjudet-
    bekampningsmedel
https://tonyasumaa.wordpress.com/2013/09/30/
    varlden-konsumerar-143-miljarder-liter-olja-per-dag/

5 – The Power of a Wild Ground

Almqvist, Carl Jonas Love, Jaktslottet: Ormus och Ariman
    m.fl berättelser, Stockholm 1969
Angulo, Jaime de, *Indian Tales*, New York 1974
Baker, Nick, *ReWild: The Art of Returning to Nature*, London
    2017
Barkham, Patrick, *Badgerlands: The Twilight World of
    Britain's Most Enigmatic Animal*, London 2013
Bonniers stora verk om jordens djur, Stockholm 1996
Dröscher, Vitus B., Hur djuren upplever världen, övers.
    Roland Adlerberth, Stockholm 1969 (Original title: *Klug
    wie die Schlangen, die Erforschung der Tierseele*; English
    title: *The Mysterious Senses of Animals*)
Dugatkin, Lee Alan & Trut, Lyudmila, *How to Tame a Fox
    (and Build a Dog)*, London 2017
Fridell, Staffan & Svanberg, Ingvar, Däggdjur i svensk folklig
    tradition, Stockholm 2007
Graebner, Livet i himmel, på jord, i vatten, övers. Roland
    Adlerberth, Stockholm 1975 (Original title: *Natur – Reich
    der tausend Wunder*)
Grahame, Kenneth, Det susar i säven, övers. Signe Hallström,
    Stockholm 1949 (Original title: *The Wind in the Willows*)
Hagberg, Knut, Svenskt djurliv i mark och hävd, Stockholm
    1950
Handberg, Peter, Jag ville leva på djupet, Stockholm 2017
Heintzenberg, Felix, Nordiska nätter. Djurliv mellan skymning
    och gryning, Lund 2013
Ingelf, Jarl, Sjukvård i djurvärlden, Stockholm 2002

Lindström, Erik, Lär känna rödräven, Stockholm 1987

Lowen, James, *Badgers*, London 2016

Meijer, Eva, Djurens språk. Det hemliga samtalet i naturens
värld, övers. Johanna Hedenberg, Stockholm (Original title:
Dierentalen; English title: *Animal Languages: The Secret
Conversations of the Living World*)

Milne, Lorus J. och Margery, Människans och djurens sinnen,
övers. Svante Arvidsson, Stockholm 1965 (Original title:
The Senses of Animals and Men)

Safina, Carl, *Beyond Words: What Animals Think and Feel*,
New York 2015

Saint-Exupéry, Antoine de, Lille prinsen, övers. Gunvor Bang,
Stockholm 1973 (Original title: *Le Petit Prince*; English
title: *The Little Prince*)

Thomas, Chris D., *Inheritors of the Earth: How Nature Is
Thriving in an Age of Extinction*, New York 2017

Thoreau, Henry David, Dagboksanteckningar, övers. Peter
Handberg, Stockholm 2017 (Original title: *The Journal of
Henry David Thoreau*)

Unwin, Mike, *Foxes*, London 2015

Waal de, Frans, *Are We Smart Enough to Know How Smart
Animals Are?* New York 2016

Wohlleben, Peter, Djurens gåtfulla liv, övers. Jim Jakobsson,
Stockholm 2017 (Original title: *Das Seelenleben der Tiere*;
English title: *The Inner Life of Animals: Surprising
Observations of a Hidden World*)

ARTICLES:

Burton, Nina & Ekner, Reidar, Indianerna i USA. Ett
reportage, Ord & Bild nr 1 1976

Ekdahl, Åke, Mickel. Naturens egen supervinnare, DN 13
april 2002

Engström, Mia, »Vi är ekologiska analfabeter«, intervju med
professor Carl Folke, SvD 8 april 2014

Flores, Juan, Bläckfisk som byter färg i sömnen förtrollar, DN
    29 sept 2019

Herzberg, Nathaniel, L'homme pousse les animaux à une vie
    nocturne, Le Monde 2 juni 2018

Snaprud, Per, Möss och människor nästan lika som bär, DN 5
    dec 2002

Walker, Matthew, Sömngåtan, SvD 3 juli 2018

800 000 kostar deppiga hundar, DN 26 jan 2018

Anthropologists discover earliest cemetery in Middle East,
    Science Daily 2 Feb 2011

https://www.livescience.com/11713-prehistoric-cemetery-
    reveals-man-fox-pals.html

https//:www.natursidan.se/nyheter/vilda-djur-utgor-bara-4-av-
    alla-daggdjur-resten-ar-boskap-och-människor

https://www.newscientist.com/article/2116583-there-are-five-
    times-more-urban-foxes-in-england-than-we-thought

https://www.sciencedaily.com/realeases/2011/02/
    110202132609.htm

6 – The Guardian Tree

Ackerman, Diane, Sinnenas naturlära, övers. Margareta Eklöf,
    Stockholm 1993 (Original title: A Natural History of the
    Senses)

Aftel, Mandy, Parfym En väldoftande historia, övers.
    Margareta Eklöf, Stockholm 2003 (Original title: Essence
    and Alchemy)

Andrews, Michael, De små liven inpå livet. Upptäcktsresa på
    människans hud, övers. Nils Olof Lindgren, Stockholm
    1980 (Original title: The Life that Lives on Man)

Beering, David, The Emerald Planet, New York 2007

Buch, Walter, Daggmasken i trädgård och jordbruk, övers
    Sixten Tegelström, Göteborg 1987 (Original title: Der
    Regenwurm im Garten)

Burton, Nina, Den nya kvinnostaden. Pionjärer och glömda
   kvinnor under tvåtusen år, Stockholm 2005
Capra, Fritjof, *The Web of Life*, London 1997
Carson, Rachel, Tyst vår, övers. Roland Adlerberth, Lund
   1979 (Original title: *Silent Spring*)
Cook, Roger, *The Tree of Life: Image for the Cosmos*, London
   1974
Dennett, Daniel C., Från bakterier till Bach och tillbaka.
   Medvetandets evolution, övers. Jim Jakobsson, Stockholm
   2017 (Original title: *From Bacteria to Bach and Back*)
Dillard, Annie, *For the Time Being*, New York 2000
Edberg, Rolf, Vid trädets fot, Stockholm 1971
Graebner, Karl-Erich, Livet i himmel, på jord, i vatten, övers.
   Roland Adlerberth, Stockholm 1975 (Original title: *Natur
   – Reich der tausend Wunder*)
——Naturen – livets oändliga mångfald, övers. Roland
   Adlerberth, Stockholm 1974 (Original title: *Natur – Reich
   der tausend Wunder*)
Greenfield, Susan A., Hjärnans mysterier, övers. Nils-Åke
   Björkegren, Stockholm 1997 (Original title: *The Human
   Mind Explained*)
Hansson, Gunnar D, Idegransöarna, Stockholm 1994
Harari, Yuval, Homo Deus. En kort historik över
   morgondagen, övers. Joachim Retzlaff, Stockholm 2017
   (English title: *Homo Deus: A Brief History of Tomorrow*)
Henrikson, Alf & Lindahl, Edward, Asken Yggdrasil. En
   gammal gudomlig historia, Stockholm 1973
Hjort, Harriet, Blomstervandringar, Stockholm 1970
Hoffmeyer, Jesper, En snegl på vejen, Betydningens
   naturhistorie, Köpenhamn 1995
Hope Jahren, Anne, Träd, kärlek och andra växter, övers.
   Joachim Retzlaff, Stockholm 2016 (Original title: *Lab Girl*)
King, Janine m.fl., *Scents*, London 1993
Kvant, Christel, Trädets tid, Stockholm 2011

Laws, Bill, Femtio växter som ändrat historiens gång, övers.
    Lennart Engstrand & Marie Widén, Stockholm 2016
    (Original title: *Fifty Plants that Changed the Course of
    History*)

Lloyd, Christopher, *The Story of the World in 100 Species*,
    London 2016

Lovelock, James, *Gaia: A New Look at Life on Earth*, Oxford
    1979, 1995

Maeterlinck, Maurice, Blommornas intelligens, övers. Hugo
    Hultenberg, Stockholm 1910 (Original title: *L'intelligence
    des fleurs*; English title: *The Intelligence of Flowers*)

Mancuso, Stefano & Viola, Alessandra, Intelligenta växter.
    Den överraskande vetenskapen om växternas hemliga liv,
    övers. Olov Hyllienmark, Stockholm 2018 (Original title:
    Verde brillante; English title: *Brilliant Green: The
    Surprising History and Science of Plant Intelligence*)

Newman, Eric A. et al., *The Beautiful Brain*, New York 2017

Nilson, Peter, Stjärnvägar, Stockholm 1991– Ljuden från
    kosmos, Stockholm 2000

Nissen, T. Vincents, Mikroorganismerna omkring oss, övers.
    Steffen Arnmark, Stockholm 1972 (Original title:
    Mikroorganismerne omkring os)

Nordström, Henrik, Gräs, Stockholm 1990

Stigsdotter, Marit & Hertzberg, Bertil, Björk. Trädet,
    människan och naturen, Stockholm 2013

Taylor, Marianne, *401 Amazing Animal Facts*, London 2010

Thomas, Chris D., *Inheritors of the Earth: How Nature Is
    Thriving in an Age of Extinction*, New York 2017

Thomas, Lewis, Cellens liv, övers. Karl Erik Lorentz,
    fackgranskning Bo Holmberg, Stockholm 1976 (Original
    title: *The Lives of a Cell: Notes of a Biology Watcher*)

Tomkins, Peter & Bird, Christopher, *The Secret Life of Plants*,
    London 1974

Watson, Lyall, *Supernature*, London 1973

———*Heaven's Breath: A Natural History of the Wind*, London 1984

———*Jacobson's Organ and the Remarkable Nature of Smell*, London 2000

Went, Frits W., Växterna, övers. Roland Adlerberth, Stockholm 1964 (Original title: *The Plants*)

Wilson, E. O., *Half-Earth: Our Planet's Fight for Life*, New York 2016

Wohlleben, Peter, Trädens hemliga liv, övers. Jim Jakobsson, Stockholm 2016 (Original title: *Das geheime Leben der Bäume*; English title: *The Hidden Life of Trees*)

———Naturens dolda nätverk, övers. Jim Jakobsson, Stockholm 2017 (Original title: *Das geheime Netzwerk der Natur*; English title: *The Secret Network of Nature*)

Yong, Ed, *I Contain Multitudes: The Microbes Within Us and a Grander View of Life*, New York 2016

ARTICLES:

Ajanki, Tord, Fattig munk blev genetikens fader, Populär historia 1/1998

Bojs, Karin Världens äldsta bacill kan förökas, DN 19 okt 2000 – Du är mer bakterie än människa, DN 17 jan 2012

Dahlgren, Eva F, Bakterier som släcker solen, DN 31 okt 1999

Ennart, Henrik, Bajsbanken kan bli framtidens föryngringskur, SvD 12 feb 2017

Forskare: så dog urtidsmänniskan Lucy, Expressen 29 aug 2016

Fredrikzon, Johan, Fotot som blev hela mänsklighetens selfie, SvD 16 sept 2017

Gyllander, Roland, Bakterien outrotlig, DN 23 okt 1994

Johansson, Roland, Antalet arter på jorden är lagbundet, SvD 20 dec 2012

Majsplantor pratar med varandra under jord, TT, Aftonbladet 4 maj 2018

Mathlein, Anders, Kaffets symbolvärde en smakrik historia,
    SvD 14 okt 2011

Niklasson, Sten, Bakterierna behöver oss – därför finns vi,
    SvD 24 jan 2013

Rydén, Rolf, Träd och människor – myt och verklighet,
    Naturvetaren nr 5 & 11 2002

Sempler, Kajanders, Munken och ärtorna avslöjade ärftlighet,
    Ny Teknik 17 juni 2017

Snaprud, Per, En formel för medvetandet, Forskning &
    Framsteg 1/2017

Spross, Åke, Bakterier ofta bättre än sitt rykte, Apoteket 3/00

https://earthobservatory.nasa.gov/features/Lawn

https://grist.org/article/lawns-are-the-no-1-agricultural-crop-
    in-america-they-need-to-die/

https://www.earthwormwatch.org/blogs/darwins-worms

https://www.forskning.se/2017/07/14/bakterier-visar-
    flockbeteende/

https://www.forskning.se/2018/08/01/livet-i-jorden-ett-
    konstant-krig-om-naring/

https://www.forskning.se/2017/12/05/livet-under-markytan-i-
    direktsandning/

https://www.forskning.se/2017/09/28/vaxter-taligare-i-
    symbios-med-svamp

https://www.forskning.se/2017/02/15/hoppstjartarnas-
    mangfald-har-sin-forklaring

https//www.slu.se/ew-nyheter/2019/1/trangsel-far-majsen-att-
    aktivera-forsvaret-och-doftsignaler-far-plantor-pa-hall-att-
    gora-likadant

https://www.svt.se/nyheter/vetenskap/8-7-miljoner-arter-pa-
    jorden